Nondestructive Evaluation of Adhesive Bonds

Using 20 MHz and 25 kHz
Ultrasonic Frequencies on
Metal and Polymer
Assemblies

by

Gilbert B. Chapman II

AuthorHouse™
1663 Liberty Drive
Bloomington, IN 47403
www.authorhouse.com
Phone: 1-800-839-8640

Published by AuthorHouse 07/11/2014

ISBN: 978-1-4969-2553-4 (sc)
ISBN: 978-1-4969-2554-1 (e)

Library of Congress Control Number: 2014912498

Any people depicted in stock imagery provided by Thinkstock are models,
and such images are being used for illustrative purposes only.
Certain stock imagery © Thinkstock.

This book is printed on acid-free paper.

authorHOUSE®

About the Author

Gilbert B. Chapman was a Senior Manager, Advanced Materials at DaimlerChrysler Corporation where he led the development and selection of advanced materials and other technologies to support advanced concept vehicle development. He was also responsible for locating and arranging for the development and transfer of advanced technologies from universities, R&D laboratories and other sources, into corporate industrial applications for product and process improvements.

Previously, Gil was responsible for developing and providing state-of-the-art nondestructive evaluation (NDE) and quality systems technologies to support corporate quality objectives for advanced materials and processes in automotive applications and, for an interim period, he supervised four Chrysler testing laboratories. His background in supporting quality with NDE technology includes serving as a Project Leader at Ford Motor Company where he performed research to develop and implement production compatible NDE technology for quality assurance of thermoplastics and composite components and processes.

Before joining Ford's Research Staff, Gil was employed at NASA-Lewis (now Glenn) Research Center developing computer automated emission spectrochemical analysis methods to support aerospace materials and propulsion research and development programs. Earlier assignments at NASA-Lewis included R&D in supersonic propulsion and high-energy fuels. He also served as NASA an occasional tour leader and recruiter.

Gilbert has a BS in mathematics and chemistry, an MS in physics, an MBA in Advanced Management from Michigan State University and a Ph.D. in physics from the University of Windsor.

Gil has been a member of ACS, APS, ASC, ASM, ASNT, CAP, ESD, NTA, SAE, SAS, SME, SPE. AAAS, IEEE and ASTM, and has held offices in several technical societies, including the American Society for Nondestructive Testing (ASNT) in which he was certified at the highest level, Level III, in six test methods. He has served as ASNT Detroit Section Chairman and was elected an ASNT Fellow. He chaired the local chapters of the Society of Physics Students and the Physics Honor Society, Sigma Pi Sigma at Kent State University. . He has organized and chaired several conferences and conference sessions on advanced materials, quality and automotive composites, has taught college classes part-time for five years, chaired a university industrial advisory board and served on several University Industrial Advisory Committees, as well as the Mt. Vernon, Ohio Academy Board.

Gil has authored or coauthored about 70 technical reports and publications, including three book chapters, on subjects such as supersonic propulsion, high-energy fuels, emission spectrochemical analysis, computer automation, nondestructive evaluation of composites, artificial intelligence, composites quality, advanced materials in concept vehicles and the implementation of technical innovations in manufacturing systems, processes and products. Gil has given about 105 presentations at Conferences and Colloquia, and organized and coordinated the first Chrysler Tech Expo, the Chrysler "Green Team" and a Polymer Composites Technology Resources Team. He also chaired the Materials Quality Improvement Team and taught Quality Improvement at Chrysler.

Biographical sketches have been published in four *Who's Who* publications and featured in *Ebony's* "Speaking of People". Gil has received recognition awards for best paper, outstanding presentations, patents, community service and for "Outstanding NASA Employees Engaged in EEO Activities", and was also recognized as the "Black Engineer of the Year for Career Achievement" and was also awarded the MLK Visiting Professor of Physics at Wayne State University.

Preface

Demands for improvements in aerospace and automotive energy-efficiency, performance, corrosion resistance, body stiffness and style have increased the use of adhesive bonds to help meet those demands by providing joining technology that accommodates a wider variety of materials and design options. However, the history of adhesive bond performance clearly indicates the need for a robust method of assuring the existence of the required consistent level of adhesive bond integrity in every bonded region. This investigation seeks to meet that need by the development of new, complementary ultrasonic techniques for the evaluation of these bonds, and thus provide improvements over previous methods by extending the range of resolution, speed and applications.

The development of a 20 MHz pulse-echo method for nondestructive evaluation of adhesive bonds will accomplish the assessment of bond joints with adhesive as thin as 0.1 mm. This new method advances the state of the art by providing a high-resolution, phase-sensitive procedure that identifies the bond state at each interface of the adhesive with the substrate(s), by the acquisition and analysis of acoustic echoes reflected from interfaces between layers with large acoustic impedance mismatch.

A low-frequency, duel-transducer approach is necessary for NDE of adhesive bonds in polymer assemblies, because interface echo amplitudes are marginal when the acoustic impedance of the substrate is close to that of the adhesive; therefore a 25 kHz Lamb wave technique was developed to be employed in such cases.

Modeling the ultrasonic echoes and Lamb-wave signals was accomplished using mathematical expressions developed from the physics of acoustic transmission, attenuation and reflection in layered media. The models were validated by experimental results from a variety of bond joint materials, geometries and conditions, thereby confirming the validity of the methodology used for extracting interpretations from the phase-sensitive indications, as well as identifying the range and limits of applications.

Results from the application of both methodologies to laboratory specimens and to samples from production operations are reported herein, and show that bond-joint integrity can be evaluated effectively over the range of materials and geometries addressed.

Acknowledgements

Grateful acknowledgements are expressed for the many people whose mentoring and management at NASA- Lewis (now Glenn) Research Center, Ford Motor Company Research and DaimlerChrysler (now Chrysler) Corporation provided advice, encouragement, financial support and guidance. NASA's William A. Gordon, and Judson Graab, Ford's Emmanuel P. Papadakis and Charles E. Feltner, DaimlerChrysler's vice president Thomas S. Moore, Thomas Asmus, Subi Dinda, and Chairman Dieter Zetsche are but a few of many in the engineering research and development environment of government and industry where much of my technical experience was obtained.

I am also grateful for the opportunity, mentoring, assistance and encouragement provided in academia, during my time in the University of Windsor's Physics Department, by Professors Roman Gr. Maev, Elena Maeva, Gordon W. F. Drake, Edward N. Glass as well as research associates Inna Severina and Sergey Titov. Dr. Ina Severina was also the principal investigator at the University of Windsor, conducting the research on degree of adhesive cure reported in chapter 6.

I am thankful for the supportive companionship of my wife, Betty, who has been an indefatigable source of encouragement and a champion of unwavering optimism. My late wife, Loretta, who while being an excellent mother of our seven children, was also active in driving all nine of us to higher levels of service, achievement and contribution.

Thank God for His many gifts: His Son, children, family, and the many people who have helped in many ways.. I hope that having received these blessings, I will "Go and do likewise".

Readers are invited and encouraged to submit comments, corrections and suggestions to the author at GBChapman2@aol.com, or call either 248 324-5037 or cell phone 248 701-0542.

Contents

List of Tables

List of Figures

1. Introduction

Advances in materials and processes technologies, along with the expanding applications of these technologies, require concomitant advances in supporting materials-characterization technologies that are developed for each specific material, processes and application, or group of related ones, in order to insure that their integrity and effectiveness are an optimal fit for each specific use. These materials-characterization technologies are essential to assure that design intent is met and that an optimal fit is accomplished for each design, material and processes application. This is especially true when these technologies are applied for supporting quality in emerging materials and processes.

Physics has proven to be a significant source of knowledge used to develop these technologies, and the contributions from physics and related physical sciences continue to build the knowledge base. Research in these sciences is an essential source of knowledge in the development of new materials and processes used to build engineering structures, and such research is also a significant source of knowledge to support the development of a variety of materials-characterization technologies for evaluating the characteristics and integrity of structures and processes before, during or after use. The research reported herein is an example of another contribution to this knowledge. It allows for a wider usage of the adhesive joining process by supporting quality and reliability improvements in adhesive bonding; thus providing an essential element in vehicle weight reduction for improved fuel efficiency.

1.1 The importance of adhesive bonding:

● Adhesive bonding provides a method of joining dissimilar materials, and thus allows vehicle designers and manufacturers wider options and more flexible choices in materials and material combinations to optimize the design and vehicle performance,

● Adhesive bonding helps meet the need for vehicle weight reduction to improved fuel efficiency. The use of advanced designs, materials and processes is increasing in order to accomplish weight reduction for improved fuel efficiency. The increased use of these materials and concomitant processes in transportation vehicles has resulted in more reliance on advanced joining technologies to provide enhancement in the design, assembly and performance of current components and vehicles. Among these advanced joining technologies, the use of adhesive bonding has rapidly expanded. This expansion, although facilitated by advances in technology, is motivated mainly by the unrelenting drive to reduce weight in order to conserve energy. Moreover, the use of adhesive bonds to improve such thickness-related characteristics as weld corrosion, vehicle stiffness, joint load distribution, vehicle noise and vibration, makes this joining technology even more attractive, as material thickness is minimized to reduce weight.

● Adhesive bonding helps to improve quality in such areas as, corrosion resistance, noise reduction, vibration dampening, fatigue life extension, improved body stiffness and more flexible styling, all without compromising the vehicle capacity, appearance and affordability to which customers have become accustomed. The vehicle weight vs. fuel efficiency data presented in Fig. 1-1 shows the exponential relationship, confirming that weight reduction in the vehicle body can allow weight reduction in other vehicle components, such as the chassis, power train and so on. Therefore the use of adhesive bonding as a structural joining method is expanding to support these advancements in materials and joining technologies. The growth of adhesive bonding in fields such as aerospace, automotive, construction, infrastructure, medical, packaging, sporting equipment and other applications has been significant, and the demand for adhesives is expected to continue to increase [1].

● The expansion of adhesive bonding as a method of joining in vehicle body structures has also grown as a result of the development and implementation of more improved adhesive bonding technologies that lead to a more wide-spread use in rapid manufacturing and assembly processes. Furthermore, other rapid joining methods, such as resistance spot welds and mechanical fasteners, are augmented by the use of adhesive bonds that provide enhanced load distribution in the joint, leading to a longer fatigue life and improved stiffness, noise and vibration characteristics of the component or assembly. Adhesive bonding, with its enhanced load distribution, also allows the use of thinner sheet metals and lighter body materials such as aluminum, polymer composites and plastics, as these less-dense materials are implemented to accomplish weight reduction.

The adhesive bond NDE research reported herein is vital to fulfilling the materials characterization requirements needed to support the technologies required to meet the needs imposed by the continuing development and application of high-performance, lightweight, low-cost materials, and the processes associated with producing, fabricating and joining them in a manner to accomplish weight reduction, without sacrificing other desirable characteristics. The following discussion will elucidate how NDE developed in the research reported here utilize the excitation, propagation, reflection, detection and interpretation of acoustic energy signals for evaluating adhesive bonds that join metals, composites and plastics into assemblies.

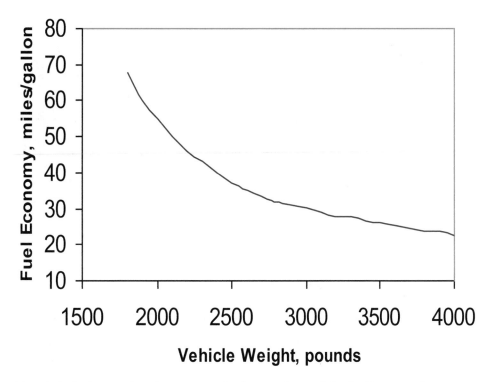

Fig. 1-1. Relationship between vehicle weight and fuel efficiency for a variety of conventional, non-hybrid, North American vehicles

1.2 The need for nondestructive evaluation to improve adhesive bonds

Because of these advantages, the growth of adhesive bonding as an automotive joining methodology continues, while the reliability of adhesive bond joints in automotive assemblies remains often inconsistent. An example of these wide variations in adhesive bond strength along a typical bond line is shown in Fig. 1-2. This example is from a bond-joint segment that is deemed to have acceptable bonding, with a bond merit factor of 0.79 [2, 3, 4]. This adhesive bond joint test data was acquired from lap-bonded specimens, tested in shear by tension loading, as schematically illustrated in Fig. 1-3, according to American Society for Testing Materials (ASTM) Standards [5, 6]. Some sample sets have yielded results with coefficients of variation (CV) as high as 0.68 [2, 7]. Such wide variations in bond-joint performance are unacceptable and would indicate that the bonding process is incapable of meeting the quality control and process capability requirements for manufacturing. However, these wide variations are usually observed after the adhesive bonding process has been approved for production launch, usually following far more acceptably consistent test results for bond-joint strength, with CV values near 0.10 or less.

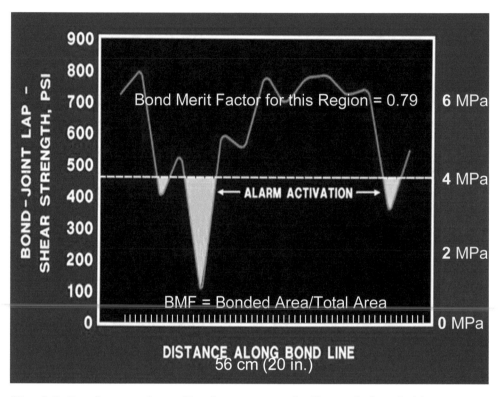

Fig. 1-2. Bond-strength profile along a typical adhesively bonded lap joint.

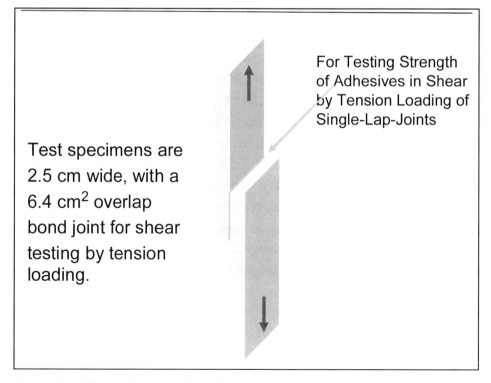

Fig. 1-3. Adhesively bonded single-lap joint for testing in shear by tension loading

These significantly large and unacceptable variations in bond-joint strength that occur after the production process has been launched, and after process control and capability is established, occur because adhesive bond quality is subsequently impacted by many factors with variability not precisely controlled in the manufacturing environment. In fact, the variability of many of these factors cannot be controlled in the manufacturing assembly facility where the adhesive bonding process is performed, because they are determined in the supply chain where the processes are performed before entering the domain of the manufacturing assembly facility. An incomplete list of 16 such factors is shown here as an example:

1. Adhesive-adherend (substrate) compatibility

2. Chemical state of the adhesive material

3. Mixing of the components of the adhesive material

4. Substrate surface condition (chemical, morphological, wetting)

5. Pre-through-post application environment (temperature, humidity, dust, etc.)

6. Application of the adhesive material for full bond coverage, quantity and location

7. Inter-penetration of adhesive into the adherend, wetting

8. Mechanical bonding of adhesive to the adherend

9. Molecular bonding of adhesive to the adherend

10. Adhesion strength between adhesive and adherend

11. Coordination, or fit, between mating contours

12. Adhesive curing conditions: temperature, pressure, time, chemistry, relative movement

13. Constant clamping force to compensate for adhesive shrinkage during post-cure cooling

14. Adhesion and/or cohesion of adherend surface layer to its sub-surface substrate

15. Tensile and shear cohesion within the adhesive layer

16. Subsequent assembly and painting process with thermal and mechanical variations

The level of sensitivity of the effectiveness of the bonding process to most of these factors is not completely known, but a significant fraction of these factors are known to vary in a typical pre-production automotive supply chain, as well as in the manufacturing assembly environment, and each factor, like links in a chain, can adversely affect the adhesive bond quality.

Because of these well-known and often-encountered variations, the adhesive bonding process is frequently monitored during production by bonding flat test specimens for lap-shear testing in tension loading, as shown in Fig. 1-3. These ideal flat specimens, cut, bonded and cured for laboratory mechanical testing, are produced under better controls than the more complex, thermally massive assemblies, and are virtually without residual stresses induced by curvature and concatenation. They therefore cannot correctly represent the bonding problems encountered in the production of actual parts, where the coordination and fit of complex-contoured components poses a serious challenge to consistent adhesive bond quality. For example, when the vehicle body structure constrains the adherend sheets, they cannot accommodate the shrinkage of the adhesive layer that occurs upon curing because of volume reduction during polymerization and thermal contraction. Moreover, the flat specimens cannot capture nor represent the often-encountered problem of spring back that can pull unconstrained bonds apart before and during curing.

Frequently, the results from these routine quality-control mechanical tests reveal the bond-quality problems long after the flawed process has been allowed to produce an abundance of poorly bonded parts that have passed through subsequent manufacturing operations and on into the production output. These assemblies must then be found and quarantined for repair or disposal. Moreover, the cost and waste associated with the value added to flawed assemblies by subsequent manufacturing operations on inferior components, and a marginally effective bond repair procedure, must also be considered. Therefore, these issues must be considered when mechanical tests are sometimes considered as candidates for monitoring adhesive bond-joint quality during manufacturing. Although mechanical bond tests are effectively utilized on specifically designed specimens during testing to establish the statistical process control and capability of the adhesive bonding process, these tests cannot be performed cost-effectively and routinely on actual parts and assemblies that have complex configurations or geometric features that do not lend themselves to such testing.

Mechanical tests, even proof tests that do not test to failure, usually destroy, deform, or deteriorate the components, making them unfit or marginally compromised for subsequent service. Mechanical tests are very valuable, however, in establishing much-needed correlations between NDE data and the mechanical performance of the bond joint.

These issues expose the adhesive bonding process as one that is not yet robust enough to be considered reliable, without the support of an effective, on-site NDE methodology to assure that the necessary bond quality consistently meets the performance requirements of the assembly. Indeed, the history of adhesive bond performance strongly indicates the need for a robust method of assuring that the required consistent level of adhesive bond integrity exists in every bonded region. Therefore, until acceptable process control and process capability are demonstrated over a sustained period, on actual production parts, NDE must be an essential component of any plan to implement adhesive bonding as a primary or secondary joining methodology in production, because in many such automotive applications, the performance, service life, and appearance of bonded assemblies are, to a significant degree, dependent on the integrity of their adhesive bonds. Furthermore, the need for NDE methods to assure the integrity of adhesive bonds intensifies as the use of

6

adhesive bonds in structural joining applications continues to expand. Therefore a method of inspecting bonds for defects is necessary to assure the desired durability and quality of these products.

NDE is also recognized as an essential asset in facilitating the establishment of process control and process capability when new materials and/or processes are implemented in production. It is an effective, efficient and often essential technology for (1) product development, (2) maintaining production process optimization, (3) assuring product quality, (4) evaluating in-service damage or deterioration, and (5) determining repair effectiveness. While the need for NDE is evident in a wide spectrum of applications that are distributed throughout these five stages of the vehicle development cycle, production and service life, the NDE method selected for each application must be specific for that application and stage.

1.3 Types of adhesive bond-joint defects to be detected

The specific causes and kinds of adhesive bond flaws rank high among those specific chemical, physical and mechanical factors that must be identified and understood about adhesive bonding mechanisms and anomalies that impact bond-joint performance. An examination of the causes and kinds of adhesive bond flaws, and their impact on bond-joint performance is essential to the selection of an effective NDE approach that will yield indications that correlate with bond-joint performance in service. It is known that structural defects of the adhesive layer negatively influence the integrity of the adhesive bond joint and decrease the strength of the assembly. Defects such as voids, inclusions, discontinuities, delaminations, kissing unbonds, porosity, air bubbles, micro-pores, delaminations, micro-bubbles, micro-cracks, etc. resulting from improper curing of the adhesive or improper adherend surface preparation. Other anomalies such as variations in bond-joint thickness, bond-line width, bond-line location, inhomogeneities, micro cracks and micro fractures are also included.

Classification of many of these defect types can be found in the wide surveys given by Adams and co-workers [8, 9] as well as Munns and Georgiou [10]. Adams and Cawley [9] classify defects into two types according to their location: defects in the bulk of the adhesive layer and defects on the adhesive–substrate interface. Furthermore, Adams and Drinkwater [8] describe four basic types of defects in simple adhesively bonded systems: gross defects, poor adhesion, poor cohesive strength, and "kissing" bonds.

Bulk defects within the adhesive layer cause a decrease of cohesive strength, because these mechanical interferences with cohesion are detrimental to the cohesive binding, or bulk tensile properties of the adhesive layer. Low cohesive strength of the adhesive material can also result from chemical causes, such as an incomplete polymerization process in the adhesive which reduces its tensile strength. Because thermoset adhesives are usually on-site activated cross-linked polymers, this compositional defect involving no-missing-material could be caused by sub-optimal adhesive stoichiometry, insufficient mixing, or improper cure of the adhesive. These factors have critical impact on all stages and facets of adhesive performance.

Gross defects include voids, porosity, cracks, and disbonds. Voids and porosity are volume elements within the adhesive bond joint from which adhesive is missing. They may be caused by insufficient adhesive, trapped water vapor that emanated from adhesive that cured by the condensation polymerization process, or vapors that migrated into the bond line from heated substrates during curing. Voids and disbonds can also arise as a consequence of the thermal expansion, and subsequent contraction, of the adhesive bond joint concomitant with curing. As the constrained bond joint is heated, the adhesive in it expands and the excessive volume resulting from the expansion escapes from the joint. When contraction occurs upon cooling, the hardened adhesive cannot return to maintain a full bond joint; hence voids and delaminations may result. The tendency toward both these problems is common because most adhesives are thermosets that are cured by heat while undergoing condensation polymerization and releasing water vapor that contributes to void formation.

Another important cause of adhesive voids is the "spring-back" effect, which occurs when the metal sheets that are joined pull apart, or spring back, after the applied holding force is removed, but before the adhesive is cured. These "spring-back" voids can result in two different situations at the first interface, and a third when the second interface is considered. One is similar to missing adhesive, where the adhesive failed to wet the first interface, that is to say one interface is coated with adhesive while the other is not. The other situation is where adhesive is stuck to the first interface, but not to the second, or where the adhesive is stuck to both first and second interfaces, but not continuously cohered between the two adherends. Such voids generally extend over long regions and result in completely unbonded joint segments. Moreover, the adhesive layer stuck to the first and/or second interface(s) will often have an irregular, rough surface where it separated upon spring-back, and that rough surface does not provide sufficient acoustic reflection to be detected. It is acknowledged by the high attenuation that indicates a bond, but the lack of an echo from the second interface, because the ultrasound cannot travel through the void to be reflected from the far adhesive-adherend interface. The shape of these defects can differ on the indication images from different sides of the sample due to its irregular reflective surface. Voids and/or unbonds caused by "pillowing" of the sheet metal between spot welds or mechanical fasteners can manifest virtually all of the characteristics of those caused by spring-back.

Cracks in the adhesive are often caused by residual stress in the bond joint due to thermal shrinkage, applied stress that may exceed design load or mode, and/or low cohesive strength resulting from sub-optimized adhesive chemistry or deficient curing of the adhesive.

Defects at the adhesive-adherend interface result in low adhesion strength. This may be the result of a weakness at the adhesive–adherend interface or internal stresses within the adhesively bonded joints. Such defects at the adhesive–adherend interface are usually derived from practices in the bonding process that reduce the overall adhesion strength, or from chemical agents and contaminants within the adhesive, within the adherend, in the atmosphere, or the poor adhesion may be caused by pre-contact partial curing of the exposed surface of the adhesive. Poor adhesion can also be caused by the presence of low-molecular-weight contaminants just below the substrate surface, where they can migrate into the interface during or subsequent to curing.

Well-known sources of substrate surface impurities commonly encountered in the adhesive bonding of polymers are: (1) mold-release agents, (2) low-molecular-weight substances in the substrate that gradually diffuse to the adherend bond surface and thereby cause marked decreases in adhesive strength and (3) chemical species, common to polyolefin surfaces, that prevent adhesive wetting of the adherend until proper surface preparation is accomplished. Consequently, poor adhesion can also caused by improper surface preparation of the adherend. Hence, these defects also result from mechanical and chemical causes.

Disbonds usually result from poor adhesion. Although this term is sometimes used to describe all regions of the bond joint that are not bonded, it is used herein to describe only those unbonded regions where adhesive is present and completely filling the bond joint, but not adhered to either or both adjoining adherend interfaces. Disbonds can result from no wetting of the adherend surface by the adhesive, from poor adhesion at the interface or from debonding due to stresses imposed by mechanical, thermal, chemical and/or environmental factors.

The so-called "kissing unbonds" or zero-volume disbonds are areas where disbonded surfaces are in contact, but not adhered or bonded. The physical principles that govern the transmission and reflection of acoustic waves in solids, cause these acoustical NDE methods to be highly effective in detecting voids and unbonded regions in adhesive bond joints. Unbonded regions of the bond joint where adhesive is present and in contact with the adherend at both interfaces, but not bonded at one or both of them, offer an often-discussed challenge to this acoustic methodology. To address this challenge, it is necessary to clearly define terms often used in NDE to describe these two undesirable states of adhesion, or lack thereof, that can occur when the adhesive is in contact with both adherend interfaces, but provide virtually no bond, because in one case there was no wetting of the adhesive on the surface of the adherend, and in the other case, there was wetting, but no significant bonding occurred upon curing. In the latter case, the weak bond that did occur often experiences "infant mortality" as it fails upon exposure to the slightest mechanical stress, even residual stress in the bond joint can cause failure.

The "kissing unbonds" and "kissing bonds" are terms that have often been used in the adhesive bond NDE community to describe either of these two states of adhesion, without distinguishing between them. These terms will be defined and used here with a clear distinction between them, because such a distinction is a prerequisite to understanding the mechanisms that cause the two states and selecting NDE method(s) to detect either or both of them. Here, the term "kissing unbonds" will be used to describe those unbonded regions of the bond joint where adhesive is present and in contact with the adherend at both interfaces, but not bonded to one or both of them. These unbonds can be easily detected acoustically and offer no challenge to the application of the NDE approaches presented herein. On the other hand, "kissing bonds" transmit acoustic energy, although not as well as good bonds, yet they are not easily detected acoustically, until those regions are exposed to minor mechanical stresses that may convert them to "kissing unbonds" which are easily detected.

Mechanical bond-joint test data acquired in previous studies [2, 7], and further analyzed here to support this concept of three states of adhesion in bond joints, are presented in Figs. 1-4. The histogram in Fig. 1-4(a) shows the distribution of test data from 25 lap-joints tested to

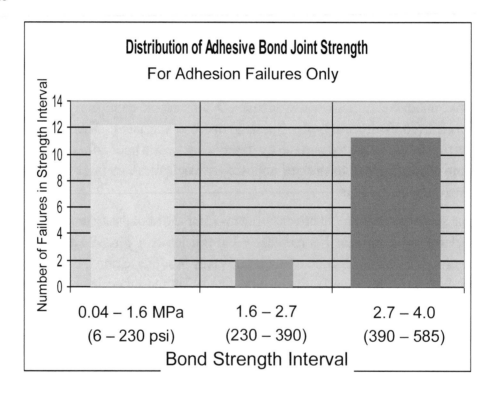

Fig. 1-4(a) Adhesion strength distribution is bi-modal for bonds that survive preparation.

failure in shear by tension loading. These failed by adhesion failure mode. The histogram in Fig. 1-4(b) shows the distribution of test data from 71 lap-joint specimens tested in shear by tension loading to failure by all modes. The data in the first histogram show a bimodal cluster of 12 values at low failure loads and 11 values at high failure loads, with only 2 of the 71 total failures occurring at loads between the two modes. The data in the second histogram show the same 12 values at low failure loads and 57 values at high failure loads. Both histograms show that only two failures occurred between these two modal clusters.

Data from the 25 adhesion failures only are identified in the second histogram, and the conclusion drawn from all failure modes shown in

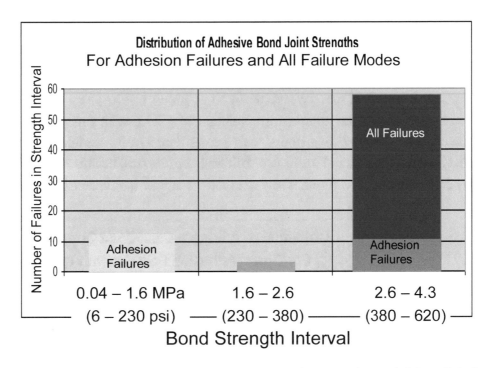

Fig. 1-4(b). Adhesive bond-joint strength distribution is bi-modal for all failure modes in fiber-reinforced plastic bond joints that survive preparation.

both histograms is the same, indicating a bimodal distribution of adhesion strengths. This conclusion is also supported by lap-shear bond-test data from bonded steel specimens, shown in the histogram plotted in Fig. 1-4(c). The source and experimental parameters of these data will be discussed further in chapter 6, where experiments that investigated the development of adhesive bond joint strength as a function of cure temperature and time are reported. These experiments also provide a better understanding of the cure chemistry and mechanisms that contribute to such bimodal distributions.

Data representing the frequency of failure at no applied load are not shown on either histogram because, in spite of the often observed existence of this zero-strength adhesion state, no such bond joints are tested, either because they are already unbonded before sample selection of the test specimens, or because they do not survived cutting during specimen preparation, and therefore cannot be tested. Including these zero-strength kissing unbonds in an assessment of adhesion states, the data would cluster near three strength levels, or adhesion states:

(1) the unbonded "kissing unbond" state, not shown here at 0 MPa (0 psi),

(2) the weak "kissing bond" state, shown here in the strength interval between 0.04 and 1.6 MPa (6 - 230 psi), and

(3) the strong bonded state, shown here in strength interval between 2.6 and 4.3 MPa (230 – 620 psi) that meets bond-joint requirements.

Note that the prevailing failure mode for all "kissing" states 1 and 2 was by adhesion failure. When bond-strength data from kissing bonds, state 2, are included in the calculation of the coefficient of variation (CV) for the distribution of strengths for all 25 adhesion failures, ranging from 0.04 MPa to 4.0 MPa (6 to 585 psi), the CV = 0.69 and the mean is 2 MPa (295 psi). When the 14 bond-strength values less than 2.6 MPa (377 psi) are excluded from the sample, the resulting CV = 0.13, with a mean of 3.4 MPa (498 psi); thus providing statistical support, with greater than 97.5 % confidence, for the hypotheses that the two samples belong to two different populations, because their means are separated by 3.11 standard deviations. Only data for the adhesion failure mode were included in this

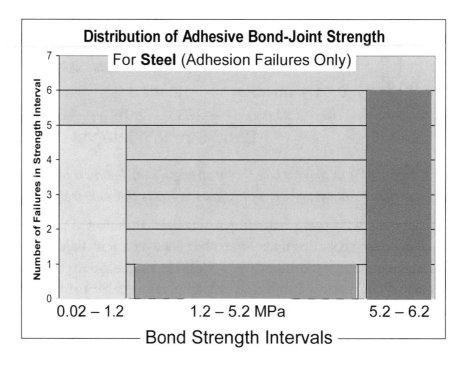

Fig. 1-4(c). Adhesive bond-joint strength distribution is bi-modal for all failure modes in steel bond joints that survive preparation.

statistical analysis, so that the results would be uncontaminated by other failure modes. Had all failure modes been included, the separation of the means would have been greater, by a much greater number of standard deviations, because

the range of all the stronger bond strength data, by all failure modes, is only 3 % greater for 32 data points instead of 11 data points for adhesion failures only; hence the separation of the populations would be proven with even greater confidence with the data that was contaminated by all failure modes.

Although these bond test data were acquired on polymer composites that were reinforced with fiberglass and bonded with a popular two-part adhesive used for such applications in the automotive industry, the observations and data supporting the argument for three bond states hold for all adhesive-adherend combinations that have been experienced thus far. Data from adhesively bonded steel specimens presented in the histogram shown in Fig. 1-4(c), also shows an obvious statistically significant separation between the weak and strong bonds. Although the variation in curing time, that produced these bond-strength variations, was consistent three-minute steps, the cure process manifested an expected dichotomy, having cured bond strengths averaging 5.8 MPa, with a CV of 0.059. Therefore, three standard deviations less than the mean strength equals 4.8 MPa, well above the population represented by the remaining data from the weak bonds.

Kissing unbonds are easily detected in adhesively bonded steel and aluminum assemblies, but kissing bonds are not, and are usually identified as bonds. State 1 can occur when the adhesive fails to wet the adherend surface upon application, because of incompatible chemistry, or when the adhesive surface to be mated to the adherend is pacified before contact with the adherend by environmental agents or by excessive open time allowing initiation of green-state cure at the exposed adhesive surface. State 2 can occur when the adhesive wets and bonds to a contaminated adherend surface, or a surface that is poorly bonded to the adherend substrate. State 2, kissing bonds, can be converted to state 1, kissing unbonds, by residual stresses concomitant with post-cure cooling of the bonded assembly. This is a familiar factor contributing to unbonds in regions where the bond joint geometry accentuates the residual stress of the bond line upon post-cure cooling.

Additional discussions of adhesive bond issues and states of bonding encountered in automotive applications have been put forth by Chapman [11]. An in depth discussion of surface preparation required for adhesion has been put fourth by Drzal, Bhurke, Rich, and Askeland [12], in which the necessity of providing appropriate surface chemistry for bonding is explained. A closer examination of the chemical and physical causes of kissing bonds and kissing unbonds will be undertaken in a later section. While a detailed understanding of the mechanisms of adhesion and the development and formation of adhesive mechanical properties in the bond joint is very important, nevertheless to date there remains no satisfying fundamental, universal understanding of the relationship between the physics, chemistry, microstructure, and the physico-mechanical properties of the materials by which adhesion to other materials is accomplished.

1.4 Selection of the preferred adhesive bond NDE approach

Several important factors must be explored and carefully considered in order to accomplish the development and selection of the optimum NDE method for a specific application. These factors include

(1) The capability for quantitative and qualitative characterization of the defect in the assembly.

(2) Consideration of the types and locations of the defects to be expected, in order to select a correct approach and to develop and implement an effective and efficient NDE methodology

(3) Consideration of the NDE method in the design, materials, production, and repair processes.

(4) Awareness of the required speed and operational simplicity to be acceptable in a mass production application.

(5) Consideration of human factors and culture of the application environment are important, because they could pose an insurmountable barrier to a successful implementation and are key contributors shaping the methodology.

The NDE approach selected in this study was guided by these five essential factors. These factors directly impact the technical approach selected, because adhesive bond mechanisms, designs and processes determine the type and geometry of the prevailing bond-joint anomalies expected, and thereby their interaction with the probing energy.

A clear understanding of the chemical, physical and mechanical mechanisms by which adhesive bonding is accomplished, as well as the chemical and mechanical stresses to which these bonds are exposed during service, will significantly facilitate the selection of an optimal nondestructive approach to effectively evaluate adhesive bond quality in the manufacturing environment in such a way that will indicate bond joint performance in service.

Research and development objectives defined by requirements and constraints that determine the requirements and constraints of the inspection methodology are essential to defining research and development objectives and focusing resources and efforts in a way that when such resources are expended, they will not only yield new scientific knowledge, but the new knowledge created can be developed into a technology that will benefit society by meeting a well-defined need. This is often referred to as "Technology Pull". Customer requirements for this research effort were established early with this approach and classified into three groups:

1. Instrument and operating procedure:

Commercial or "Turn-Key" availability to manufacturing personnel

Portability (small size or transportable)

User-friendly, operational simplicity

Minimal interference with production

Cost-effective, robust and reliable usage in manufacturing environment

2. Bond performance indicators:

Local bond integrity (LBI) index, measured over 1 to 3 cm [2, 7]

Bond merit factor (BMF) [2, 7], determined for a region of about 40 to 50 cm,   Graphic Display of bonded and/or unbonded regions

3. Correlation of NDE data with joint performance:

High correlation of NDE indicators with bond joint performance, as measured by mechanical tests and actual in-service performance

These requirements must be reviewed and revised, as often as needed, in discussions between researcher and customer.

These valuable interactions with the customer, and a review of NDE technologies, have resulted in complementary ultrasonic methods for nondestructive evaluation (NDE) of adhesive bonds. These methods have been selected for development in order to provide the necessary effective and efficient bond quality assurance methodologies for an expected variety of manufacturing application requirements. These methods provide improvements over previous methods implemented in production, and cover the range of bond evaluation situations, with the required capability and effectiveness for detection resolution and inspection speed, as these requirements vary with the application.

1.5 Background summary of acoustic adhesive bond NDE techniques

Having acknowledged the many advantages that adhesive bonds offer in joining vehicle components and assemblies, while also raising awareness of their limitations in providing consistent bond joint quality, it is understandable why the implementation of structural adhesive bonding is to be accompanied by NDE methodology for assuring bond-joint quality. These methods are many and vary over a wide range of technical approaches and application strategies. Although this background summary will focus on acoustic NDE techniques, it is valuable to include an evaluation of the effectiveness and efficient of other NDE approaches, so that the acoustic approach adopted for this research will not have resulted from a search that was blindly restricted by a narrowly defined arbitrary acoustic paradigm.

These broader bond NDE techniques range from the early coin-tap test, where a well-tuned, experienced ear listened to and analyzed the sound resulting from tapping the joints of a bonded assembly, to the more sophisticated procedures where the acoustic vibrations or thermal energy excited in the joint, and transmitted or reflected by it, get received and analyzed electronically, and user-friendly results displayed. In the recent past, much progress has been made in the development and improvement of these methods for the interrogation of adhesively bonded structures.

Among the NDE techniques studied, acoustic techniques are primary effective tools for nondestructive evaluation of adhesive bonds, because these methods derive their effectiveness from the propagation of mechanical stress waves whose propagation characteristics are closely related to the mechanical properties of the materials and interfaces through which they are propagated and/or reflected. Thus they provide for the detection of voids, delaminations, porosity, cracks, missing adhesive and lack of adhesion in the bond joints, as these anomalies adversely affect the mechanical performance of the joint. Acoustic interrogation can also detect the degree of cure within the adhesive, because the modulus of the adhesive is influenced by its degree of cure, and hence an effect on the velocity and attenuation of the acoustic energy. The potential for obtaining a great deal of information by the application of acoustic interrogation methods to the investigation of adhesive bond joints explains why these methods have experienced increasing attention given to them by a wide range of researchers.

Acoustical techniques allow for the measurement of not only important quantitative material parameters such as the elastic modulus and mechanical energy loss, but also provide data from which qualitative material characteristics can be obtained. Therefore, it is highly likely that accurate information about the locations and types of structural anomalies, different types of defects and their distributions can be detected, as well as the identification of their internal structures. This would apply to any adhesively bonded system, regardless of the nature of adhesive and adherend, in which the adhesive must provide the joint integrity for the whole multilayered assembly. On the other hand, other methods may be less effective. For example, moderate thermal transmission may exist across an interface where intimate thermal contact occurs, as in "kissing unbonds", but no molecular or microscopic mechanical coupling exists to transmit acoustic energy or provide mechanical bonding in service.

The most common acoustic techniques, such as ultrasonic scans, resonant ultrasonic spectroscopy, and Lamb-wave methods have been reviewed by E. Maeva, et al [13], in which an analysis of typical defects that can occur in adhesive joints was undertaken, along with an examination of their causes and methods of detection. The findings and supporting theory were presented and discussed in the review. The progress of the study of adhesion mechanisms and the role of the interfacial properties and surface conditions in the adhesion process are also discussed therein.

1.6 Scope of research reported

The NDE methods reported herein are those that resulted from research to support the development of NDE technology that can be implemented in a manufacturing environment. Hence the research focused on ultrasonic NDE methods that can be implemented in a way that meets the requirements of the mass-production environment concomitant with automotive manufacturing and meet the quality and productivity needs of the automotive industry, while satisfying the constraints of the business enterprise, especially during the implementation of new or improved designs, materials and processes.

These methods are:

1. A high-frequency ultrasonic pulse-echo method - The development of a 20 MHz pulse-echo method for NDE of adhesive bonds is reported first. It allows the assessment of bond joints with adhesive as thin as 0.1 mm. This new method advances the state of the art by providing a high-resolution procedure for in-plant assurance of bond integrity in regions with narrow inspection access. The inspection procedure, resulting indications, the physics of their origins, and the methodology for extracting interpretations from the indications will be presented to show how the presence of bonds at the first interface, between the first metal layer and the adhesive, are recognized by the increased attenuation rate of echoes reverberating in the first metal sheet, and by echoes from the second adhesive interface. Bond integrity at the second interface is evaluated by a phase-sensitive analysis of the echoes reflected from that adhesive-metal interface. Application of this NDE methodologies to laboratory specimens and to samples from production operations, as well as in the production facility, show that joint integrity at both interfaces can be robustly evaluated by the 20 MHz pulse-echo method, using a 3-mm transducer element with a 6-mm diameter, 7-mm long standard delay line with couplant.

2. A low-frequency Lamb-wave propagation method - The high resolution of the 20 MHz pulse-echo method complements this low-frequency 25 kHz Lamb-wave method that provides higher inspection speed, while sacrificing detection resolution and information about which bond-joint interface contains the unbonded region. This low-frequency method requires wider bond-joint access, but is effective where bond joints materials do not offer a large acoustical impedance mismatch at interfaces between joined adhesive-adherend layers. This gives an advantage to the low-frequency 25 kHz Lamb-wave method when NDE of bonds in polymer composite or plastic assemblies is required, because these materials have acoustic impedances that are close to those of the adhesive. Furthermore, this technique can also be an effective NDE method for joints in plastic assemblies joined by welding, joints in which acoustic impedance mismatch does not exit.

The development of these two novel approaches to rapid and reliable evaluation of a variety of adhesive/adherend materials and assemblies for their mechanical and physical properties is reported, along with a discussion of the needs motivating the development and the physical principles guiding the selection of each technical approach and forming the foundation on which the technique effectively functions. The elucidation of physical principles is included so that it can be seen how the physics determines the fit of the NDE technology to the need and application. Hence, the discussion will provide insight onto the principles of physics that under gird each technology, and will supply sufficient technical details to foster the implementation of each NDE technique in other appropriate applications that may extend well beyond the examples reported herein. Moreover, the report will seek to invite and encourage further research to develop additional NDE technologies to meet the continually emerging needs in materials research, process development and commercial production.

This research is a part of the development a larger comprehensive quality system that will utilize NDE technology as one of its components to provide early feedback of quality information to the process, and help advance adhesive bonding technology to a level of materials and process reliability where inspection will no longer be necessary.

2. Review of Literature, Models and Mechanisms

In adhesively bonded structures, empirical relationships between the characteristics of the structural micro-defects and the mechanical properties of the assembly are often used, but the coefficients of correlation between these two groups of parameters, obtained using different approaches and methods, including morphological and mechanical characterization, are comparatively low. This problem is partly caused by the impossibility of conducting both mechanical tests and microstructure evaluation on the same specimen due to the destructive nature of conventional mechanical test methods.

Acoustic methods have been widely used for NDE of adhesively bonded metal, plastics and composite components and assemblies. Much progress has been made during the past two decades in the development and improvement of acoustic methods for the evaluation of adhesively bonded structures. These methods allow us to detect voids, delaminations, porosities, cracks, and poor adhesion in the bond joints these assemblies. This review will focus on the most common techniques, such as normal and oblique ultrasonic scans, resonant ultrasonic spectroscopy, and Lamb-wave methods. Typical defects that can occur in adhesive bond joints were identified and discussed in the introduction, in order to define and describe the variety of defects requiring NDE. This review will then focus on the ultrasonic NDE technologies that may provide effective approaches to evaluating adhesive bond joints for these defects. An analytical review will be done here, along with the causes of defects that are related to the bonding process. The progress of the study of adhesion mechanisms and the role of interfacial properties and surface conditions in the adhesion process are also reviewed.

The aim of the review of the literature provided in this chapter is to evaluate the effectiveness of the acoustic methods that have been developed over the past few decades for applications that focus on detecting bonded joint defects identified and discussed in the introduction. The basic theories and models of adhesively bonded structures are then described, along with the general mechanisms of adhesion. Moreover, because the effectiveness of these approaches are supported by the fundamentals of physics, further pursuit of these and other acoustic approaches have high potential for yielding improvements in resolution, broader applications and additional insights into adhesive bond quality. Finally, the review concludes by analyzing the most important developments in the normal and oblique incidence techniques, resonance spectroscopy, Lamb-wave methods, and theory of the nonlinearity of ultrasound through adhesively bonded joints.

2.1 Acoustic waves in visco-elastic materials

A lowest order approximation of the restoring forces on an acoustic wave propagating in an elastic medium can be given by Hooke's law. In 1867, Maxwell [18] showed that Hooke's law is insufficient for materials that are neither ideal elastic solids nor Newtonian liquids. This includes virtually all engineering materials that manifest both elastic and viscous properties. The adhesive joints examined in this study are examples of such viscoelastic materials. Moreover, rigorous relationships exist between the physical and mechanical properties of materials and their chemical composition, their molecular, morphological and geometric structure, as well as the way they are combined in design and application. Joining similar or dissimilar materials by adhesive bonding involves the use of viscoelastic materials in engineering applications, where binding forces that are in general nonlinear, can cause a nonlinear modulation of reflected or transmitted waves in the ultrasonic frequency range. As a result, the higher harmonics, which can be generated by the excitation of an interface by a mono-frequency wave, allow information to be obtained about the properties and structure of the adhesion joining that interface.

The great variety of acoustical methods can provide many possibilities for investigation under various scenarios. During the last few decades, a stable set of acoustical methods have been developed and a review of early work in this field has been given by Thompson and Thompson [19]. Three principal types of methods exist for the characterization of adhesion phenomena by ultrasonic methods:

- The classical pulse-echo [20–22] and through-transmission methods are applicable for locating internal defects in bonded structures, including flaws, voids, and internal cracks. Both transmission and reflection modes are used. The transmission technique is effective in detecting uncured zones but is often not feasible in practice because it requires access to the both sides of the specimen simultaneously, whereas the pulse-echo approach requires access to only one side.

- A second group of methods is based on the phenomenon of acoustic resonance [23], where the pulse is considered as a broadband excitation for the various resonance frequencies of the system. This method has been used to detect cracks in graphitized carbon electrodes [24] and also to observe the absence of bonding in the ceramic tiles used as thermal insulation on space shuttles [25].

- The third group of methods for the investigation of adhesively bonded joints is based on surface, plate, and interface wave propagation. These types of waves are sensitive to the mechanical properties of the adhesive, combined with that of the bonded substrate material(s), [26] and to the boundary conditions between the adhesive and substrate [27].

Common techniques, such as normal and oblique ultrasonic scans, resonant ultrasonic spectroscopy, and Lamb-wave methods are discussed in this review. Analysis of the typical defects that can occur in adhesive joints and their causes are presented. Figure 2-1 is a schematic illustration of the longitudinal cross-section of such a bond joint under inspection by a transducer with a delay line, and showing the adhesive and substrates. The interfaces between the adhesive and the substrate layers are identified on the left side of the figure as interfaces 0, 1, 2 and 3, the identifying numbers that will be used throughout this chapter ant those that follow. The surface of the substrate that is adhered to the adhesive is often referred to as the adherend and, as discussed in the Introduction, the lack of adhesion can still exist

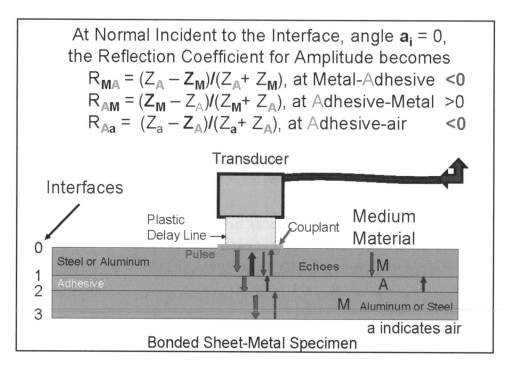

Fig. 2-1. Longitudinal cross-section of an adhesive bond joint under ultrasonic pulse-echo inspection by a transducer with a delay line.

when adhesive fills the joint. The progress of the study of the adhesion mechanism and the role of the interfacial properties and surface conditions in the adhesion process are reviewed.

2.2 Mechanisms and models of adhesively bonded joints

The study of the adhesion mechanism began in the 1920s when MacBain and Hopkins introduced the mechanical interlocking model [28]. In spite of numerous papers that have reported on the problems with adhesives made of plastic materials, fundamental knowledge about the adhesion processes is still not well developed, and no global approach or theory describes all adhesion phenomena in detail. This lack of a general theory may be explained by the fact that the formation of adhesively bonded joints involves a great number of associated and interdependent processes. The bonding of the adhesive is the sum a number of mechanical, physical, and chemical forces that overlap and influence one another.

The review by van der Leeden and Frens [29] describes four principal groups of adhesion models: mechanical theories, diffusion theories, electrostatic theories, and adsorption theories. Fourche [30] gives a similar classification, but also includes a model of weak boundary layers, which, despite some criticism leveled at it in the past [31], remains important for explaining some cases of poor adhesion. In this section, we will introduce some important adhesion models that can help to provide an understanding of the adhesive bond formation.

2.2.1 Mechanical interlocking model

The mechanical interlocking (or hooking) model is the one of the earliest adhesion theories; it was introduced by MacBain in 1925 [28]. The model involves the mechanical (physical) interlocking between irregularities of the substrate surface and the cured adhesive at the macroscopic level. Of the three types of irregularities shown in Fig. 2-2, only type *b* is shaped to provide mechanical interlocking [29]. However, when there are surface irregularities of types *a* or *c*, the strength of adhesion depends on the direction of the applied force, when only mechanical hooking is present. Mechanical interlocking is strongly affected by the roughness, porosity, and irregularities of the surface, but only under sufficient wetting

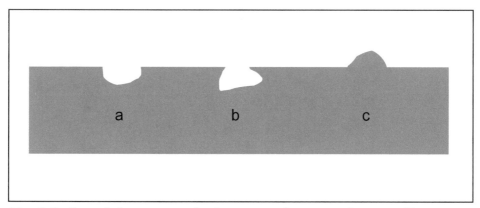

Fig. 2-2. One of three types of substrate surface irregularities offers topography to support mechanical bonding.

of the substrate by the adhesive. In fact, the lack of wetting of the surfaces of the adherend can prohibit adhesive bonds from forming at all. Even the mechanical interlocking component of bond strength is minimized without wetting, because the adhesive does not "wick" into "type b" surface features that foster mechanical interlocking. Wetting of the adherend by the adhesive is necessary, but not sufficient, to assure a good chemical bond.

Details of the wetting theory will be described in the section devoted to adsorption theory. Hence, for strong adhesion, the adhesive must not only wet the substrate but also have the proper rheological characteristics for penetrating into its pores in a reasonable time, before penetration is inhibited by the increase in viscosity that is concomitant with curing. Low adhesive viscosity promotes greater interfacial strength due to its faster and more complete penetration into the micro-voids and pores.

Examples of adherend surface topology of once-bonded surfaces of fiber-reinforced polymer composite bond joints are shown in the atomic force microscope scans of Fig. 2-3 and Fig. 2-4. Figure 2-3 shows the sizes and topography of surface irregularities for a poor bond and those for a good bond are shown in Fig. 2-4. The sizes of such surface features appear to be on the order of 0.5 μm to 5 μm, as indicated from these two micrographs.

Mechanical interlocking is a major factor in the adhesion of fibrous or porous materials such as textiles [32], wood [33], and paper as well as natural rubber to steel [31]. Later, in a number of works [34–36] the importance of physical interlocking at the microscopic level was demonstrated where the porosity or micro roughness of the surface provides a "composite-like" interphase with the adhesive. Thus, surface treatment of the polyethylene

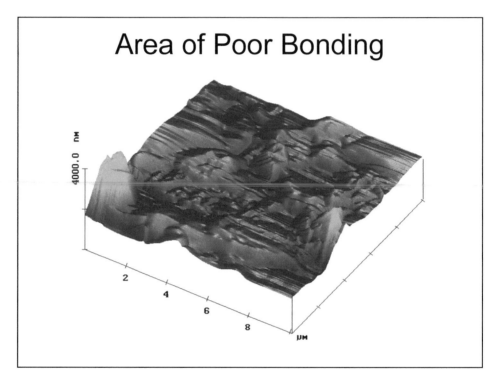

Fig. 2-3. Atomic force microscope scan of poorly bonded thermoplastic substrate surface after joint failure. Glass fiber diameter was 10 μm.

fiber with chromic acid or plasma significantly improves adhesion quality [37] due to the irregularities of the fiber's surface and increased interfacial area.

Despite its obvious appeal, the model of mechanical interlocking cannot be considered as a universal adhesion theory because good adhesion can occur even between perfectly smooth-surfaced adherends. Moreover, this theory does not consider any factors that occur on the molecular level at the adhesive/substrate interface. Mechanical interlocking should only be considered as a composite attribute in the overall view of adhesion mechanisms. This model can be effectively applied in situations where the substrates are impermeable to the adhesive and where the surface of the substrate is sufficiently rough.

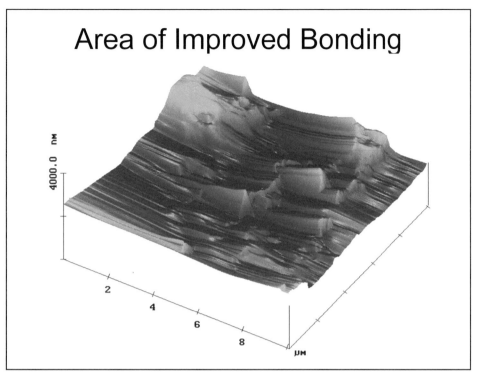

Fig. 2-4. Atomic force microscope scan of a well-bonded fiber-reinforced thermoplastic substrate surface after joint failure. Glass fiber diameter was 10 μm.

2.2.2. Diffusion theory

The diffusion theory was proposed by Voyutski [38], who explained adhesion between polymeric materials as being the result of interdiffusion of the macromolecules of the two polymeric materials at the interface as illustrated in Fig. 2-5. According to the diffusion theory both the adhesive and substrate must be polymers, which are mutually miscible and

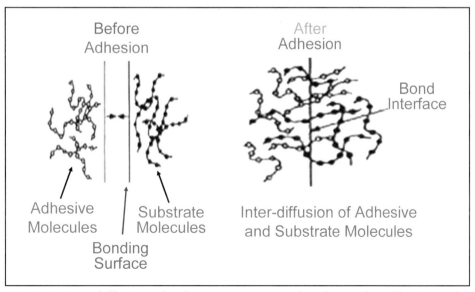

Fig. 2-5. Interdiffusion of polymeric macro-molecules at the adhesive-substrate interface can contribute to effective bonding.

23

compatible. The theory asserts that two volumes of polymers that are placed in contact under a constant assembly pressure will diffuse together following Fick's laws of diffusion. When concentration remains constant with time, steady state conditions, then F_x, flux in the x direction, is proportional to the concentration gradient, so that

$$F_x = -D\left(\frac{\partial c}{\partial x}\right)$$
(2-1)

where

D is the diffusion constant and

c is concentration.

When concentration varies with time, Fick's second law determines the diffusion constant can be used to calculate this variation with time, or the concentration variation with time and space can be used to calculate the diffusion constant by

$$\left(\frac{\partial c}{\partial t}\right) = D\left(\frac{\partial^2 c}{\partial x^2} + \frac{\partial^2 c}{\partial y^2} + \frac{\partial^2 c}{\partial z^2}\right) \quad \text{or} \quad \frac{\partial c}{\partial t} = D\nabla^2 c$$
(2-2)

According to Lee [31], the molecular diffusion constant, D, can be calculated from the Buche-Cashin-Debye equation

$$D\eta = \left(\frac{N_A \rho kT}{36}\right)\left(\frac{L_M^2}{M}\right)$$
(2-3)

where
c is concentration
η is bulk viscosity,
N_A is Avogadro's number,
ρ is density,
k is Boltzmann's constant,
T is absolute temperature
M is molecular weight and
L_M^2 is mean-square end-to-end length of a single polymer chain.

It has been shown that interdiffusion is optimal when the solubility characteristics of both polymers are equal [39, 40]. The chain length of the macromolecule, the concentration, and the temperature all have a significant influence on the mobility of the macromolecules and, therefore, on the interdiffusion process and on the adhesive strength [28].

Although increasing attention is being paid to the study of the interdiffusion process, the kinetic performance of the diffusion mechanism is still difficult to predict and not completely understood at present. Vasenin [41] developed the kinetic concept of adhesion based on Fick's first law. This quantitative model states that the amount of material diffusion in a given direction across an interface is proportional to the contact time and gradient of concentration. Later, the diffusion kinetics were rewritten in light of the reptation theory of de Gennes [42] and later extended by several authors [43, 44]. The reptation theory has made much progress in the fundamental understanding of the molecular dynamics of polymer chains and it has been applied to study the tack, green strength, healing, and welding of polymers.

It is possible to evaluate the depth of interpenetration, x, from the contact time, t, the gradient of polymer concentration parameters, and the number of macromolecular chains crossing the interface, $L_o(t)$ [31], by

$$x\ t\ \approx t^{\frac{1}{4}} N_M^{-\frac{1}{4}} \qquad (2\text{-}4)$$

$$L_0\ t\ \approx t^{\frac{3}{4}} N_M^{-\frac{1}{4}} \qquad (2\text{-}5)$$

where

N_M is number of monomers per molecular chain in the polymer.

A direct relation exists between the concentration gradient and the contact time. Vasenin studied the peel energy for joints bonded with polyisobutylenes of different molecular weights and established that peel strength is proportional to the contact time $t^{1/2}$.

Finally, however, according to the literature, the diffusion model of adhesion is not thought to contribute to adhesion if the substrate polymers are crystalline or highly cross-linked or if contact between two polymeric phases occurs far below their glass transition temperature. It has also been found to be of limited applicability if the adhesive and substrate are not soluble.

2.2.3 Electrostatic theory

The basis of the electrostatic theory of adhesion was developed by Deryaguin and Smilga [45] and is based on the difference in electro-negativities of adhesively bonded materials. According to this model, the adhesive-substrate system can be considered as a capacitor, whose plates consist of the electrical double layer, which occurs when two different materials come into contact with one another [30, 45, and 46]. The electrical double layer forms when electrons cross the interface thereby creating regions of positive and negative charge. According to the electrostatic theory [46], these electrostatic forces at the interface account for of the attraction between the adhesive and the substrate.

Hays [47] considered electrostatic adhesion of two materials (donor and acceptor) between anions and cations for two cases: (a) when the average distance r between cationic and anionic groups is smaller than the distance a between donor–acceptor pairs , $r \ll a$; and (b) when $r \gg a$.

For case when $r \ll a$, the electrostatic attractive force Fe between cation and anion pair is equal to

$$F_e = \frac{e^2}{4\pi k\varepsilon r^2}$$

(2-6)

and the adhesive force per unit area is

$$P_{r>a} = nF_e = \frac{Ne^2}{4\pi k\varepsilon r^2}$$

(2-7)

where

e is electron charge,

k is interfacial dielectric constant for the two materials,

ε is permittivity of the medium, and

N is density of the anion–cation pairs per unit area.

In the case when the distance between donor and acceptor groups is comparable to the donor-acceptor pair distance, the attractive force between cations and anions increases because of the electric field from adjacent dipoles. And for the case when $r \gg a$, the electrostatic force and adhesive force per unit area are

$$F_e = \frac{Ne^2}{2k\varepsilon_0}$$

(2-8)

$$P_{r>a} = nF_e = \frac{N^2 e^2}{2k\varepsilon_0}$$

(2-9)

Hays [47] has established that the electrical contribution to the adhesion is comparable or even greater then the van der Waals component if the density of anion–cation parts is approximately $10^{17}/m^2$ or greater.

Adhesion between two materials when one of the contacting parts has a certain radius of curvature is important in many technological processes [48]. Adhesion between these two materials has a more complicated nature. In areas of close contact, the adhesive force is similar to the one for the planar configuration. Outside this area, the total adhesion force, Fa, consists of the Lifshitz–van der Waals force, $FvdW$, the electrostatic force, Fe, as well as capillary and double-layer forces, with only the Lifshitz–van derWaals and electrostatic forces playing a major role [49]

$$F_a = F_{vdW} + F_e .$$

(2-10)

For the rigid spherical particle with radius r and uniformly distributed surface charge Q, the Coulomb electrostatic image force is [47]

$$F_e = \frac{\alpha Q^2}{16 \pi \varepsilon_r^2} ,$$ (2-11)

where

ε is permittivity of the medium surrounding the particle, and

α is a coefficient that depends on

kp, the dielectric constant of the particle.

It is assumed that a spherically symmetric charge distribution can be modeled as a single charge point, Q, in the centre of this sphere. Values of Fe computed using this model are usually lower than the corresponding measured values [50, 51]. To solve this problem, the concept of a nonuniform distribution of the charge was proposed [52]. Feng and Hays [48] note that the nonuniform distribution of surface charge and, particularly, the high charge density in the contact region enhance the magnitude of the electrostatic image force. Czarnecki and Schein [50] modeled the charge distribution as a set of point charges and introduced additional electrostatic force components due to the charges' proximity to the plane; in this case, the total electrostatic force was found to be 2.2 times higher than that calculated using (11).

The other force that makes a major contribution to adhesion is the van der Waals force, which originates from the molecular interaction due to various polarization mechanisms. For the same rigid spherical particle, the van derWaals force can be calculated as

$$F_{vdW} = \frac{Ar}{6h^2}$$ (2-12)

where

A is material-dependent Hamaker constant, and

h is the minimum distance between the two materials.

The value of the van der Waals force depends also on the surface roughness and decreases as the roughness increases.

Despite numerous studies, the role and magnitude of the electrostatic force in adhesion is still being debated; however, an increasing number of authors [48, 49, 53] have come to the conclusion that the values of both the van der Waals and the electrostatic forces can vary over a range that exceeds one order of magnitude, depending upon the particle size and surface properties (roughness and charge distribution). This could be "the primary reason for controversial results reported in the literature when the comparison conditions were not specified adequately" [48].

The significant limitation of the electrostatic model is that it is applicable only in the case of incompatible materials such as a polymer and a metal. Moreover, because the electrostatic model requires a large number of parameters for predicting adhesion processes, it has found little practical use in engineering.

2.2.4. Adsorption theory

The adsorption theory is the most generally accepted model; it was introduced by Sharpe and Schonhorn [54] and considers adhesion the result of intermolecular or interatomic forces at the interface between the adhesive and the substrate after their intimate contact. Forces between adhesive and substrate can be primary (ionic, covalent, or metallic) or secondary (van der Waals or hydrogen bond). The theory includes several models that sometimes are considered as separate theories: wetting, rheological, and chemical adhesion models.

It is well known that for good adsorption, effective wetting is essential to provide close contact between two substrates. Studies by Drzal and colleagues on surface preparation of polymers for adhesive bonding [13], as well as several comprehensive reviews [31, 55, 56], have provided major results regarding wetting and wettability of polymer surfaces for bonding.

The equilibrium balance of forces at the contact between three materials phases is given by Young's equation [57]

$$\gamma LV \cos \theta = \gamma SV - \gamma SL \tag{2-13}$$

where

γLV is the liquid-free surface energy,

γSV is the solid-free surface energy,

γSL is the solid–liquid interfacial free energy, and

θ is the contact angle between the solid–liquid interface.

Usually, surface tension and (or) interfacial tension parameters are substituted for free energy.

For wetting to occur spontaneously, the condition must exist where

$$\gamma SV \geq \gamma SL + \gamma LV \tag{2-14}$$

should apply. If the γSL is not significant, this criterion can be simplified to

$$\gamma SV \geq \gamma LV \text{ or } \gamma_{adherend} \geq \gamma_{adhesive} \tag{2-15}$$

i.e., the adhesive will spread on the substrate when the surface free energy of the substrate is greater then that of the adhesive. Poor wetting causes less contact area between the substrate and the adhesive and more stress regions at the interface and, accordingly, adhesive joint strength decreases.

The most common interaction at the adhesive–substrate interface results from van der Waals forces. These are long-range forces (effective from a distance of 10 nm), which consist of dispersion and polarization components. Van der Waals forces directly relate to fundamental thermodynamic parameters, such as the free energies of the adhesive and substrate, and allow a reversible work of adhesion of the materials to be calculated for the materials in contact. Thermodynamic aspects are described by Duprée's equation [58], which expresses the thermodynamic work of adhesion W_A as

$$W_A = \gamma LV + \gamma SV - \gamma SL \qquad (2\text{-}16)$$

Combination (11) and (14) give the Young–Duprée equation

$$W_A = \gamma LV (1 + \cos \theta) \qquad (2\text{-}17)$$

Sharpe and Schonhorn [54] have shown that the adhesive joint strength is influenced by the ability of the adhesive to spread spontaneously on the surface when the joint is initially formed.

Hamaker developed a theory for the calculation of the attraction energy between two phases 1 and 2 [59]

$$W_A^{vdW} = \Delta G_{12} = -\frac{A_{12}}{12\pi d^2} \qquad (2\text{-}18)$$

where

A_{12} is the Hamaker coefficient, and

l the distance between phases.

If the phases are solids or liquids

$$A_{12} = \frac{3}{2} kT \sum_{n=0}^{\infty} \left[\frac{\varepsilon_1(i\omega_n) - 1}{\varepsilon_1(i\omega_n) + 1} \right]\left[\frac{\varepsilon_2(i\omega_n) - 1}{\varepsilon_2(i\omega_n) + 1} \right] \qquad (2\text{-}19)$$

where

k is Boltzmann's constant,

T is temperature, and

ε_1 and ε_2 are the macroscopic dielectric susceptibilities of the phases 1 and 2 that are a function of

ω_n, the frequency and

n is the quantum number of the relevant oscillation.

Acid–base interaction is a major factor among short-range (<0.2 nm) intermolecular forces and involves hydrogen bonding, electron donor–acceptor, or electrophile–nucleophile interaction. Fowkes [60] proposed that interfacial tension γ may be expressed by two terms: a dispersive component, γ^d, and a polar component, γ^p, such that

$$\gamma = \gamma^d + \gamma^p \qquad\qquad (2\text{-}20)$$

The dispersive component is directly concerned with the Lifshitz – van der Waals interaction; while the polar component represents all the short-range nondispersive forces, including hydrogen and covalent bonds.

According to Fowkes and co-workers [60], Allara and co-workers [61], Fowkes [62] as well as Owens and Wendt [63],

$$\gamma_{12} = \gamma_1 + \gamma_2 - 2\sqrt{\gamma_1^d \gamma_2^d} - 2\sqrt{\gamma_1^p \gamma_2^p} \qquad\qquad (2\text{-}21)$$

and adhesion work for dispersive forces is

$$W_A^{LW} = W_{12}^d = 2\sqrt{\gamma_1^d \gamma_2^d} \qquad\qquad (2\text{-}22)$$

where

γ_1^d, γ_2^d are the dispersive components of the surface free energy of the substrates 1 and 2 illustrated in Fig. 2-1, while

γ_1^p and γ_2^p are the polar components.

The work of adhesion in the case of acid–base (ab) interaction W_A^{ab} is, according to Fowkes [60],

$$W_A^{ab} = -fn^{ab}\Delta H^{ab} \qquad\qquad (2\text{-}23)$$

or can be calculated by combining (17), (20), and (22)

$$W_A^{ab} = \gamma 2 \left(1 + \cos\Theta\right) - 2\sqrt{\gamma_1^d \gamma_2^d} \qquad\qquad (2\text{-}24)$$

where

f is the correction factor (to convert the heat of the interfacial acid–base interaction into free energy), which is close to unity,

n^{ab} is the number of acid–base pairs involved per unit area and

ΔH^{ab} is the enthalpy of the acid–base complex formation.

Therefore, W_A may be determined from measurements of the contact angle using groups of standard liquids with known γ^d and γ^p values [60, 63, and 64]. Fowkes' approach is currently criticized for substantial overestimation of the apolar, or dispersive forces [65]. To better describe the polar component in terms of acid–base interaction this approach was later extended by van Oss, Good, and Chaudhury [59, 66, 67]:

$$\gamma = \gamma^{LW} + \gamma^{ab} \tag{2-25}$$

where γ^{LW} is the Lifshitz – van der Waals component and

γ^{ab} is the acid–base component of the interfacial energy.

The contribution of the acid–base interaction to the interfacial energy is subdivided into two terms regarding donor functionality: the electron acceptor γ^+ and electron donor γ^- [59]. Then, the acid–base contribution to the interfacial energy can be determined

$$\gamma_{12}^{ab} = 2\left(\sqrt{\gamma_1^+} - \sqrt{\gamma_2^+}\right)\left(\sqrt{\gamma_1^-} - \sqrt{\gamma_2^-}\right) \tag{2-26}$$

If the surface involves both Lifshitz – van der Waals and acid–base interactions, the total interfacial tension between the two phases is expressed as

$$\gamma_{12} = \gamma_1 + \gamma_2 - 2\sqrt{\gamma_1^{LW}\gamma_2^{LW}} - 2\sqrt{\gamma_1^-\gamma_2^+} + \sqrt{\gamma_1^-\gamma_2^+} \tag{2-27}$$

Combination of (15), (19), and (27) yields the expression

$$\gamma(1 + \cos\theta) = 2\left(\sqrt{\gamma_1^{LW}\gamma_2^{LW}} + \sqrt{\gamma_1^-\gamma_2^+} + \sqrt{\gamma_1^+\gamma_2^-}\right) \tag{2-28}$$

Adhesion work can be calculated as

$$W_A = 2\left(\sqrt{\gamma_1^{LW}\gamma_2^{LW}} + \sqrt{\gamma_1^+\gamma_s^-}\sqrt{\gamma_l^-\gamma_s^+}\right) \tag{2-29}$$

According to the literature, two different approaches can be used to obtain information about the acid– base nature of surfaces from wetting measurements. In the first approach suggested by Fowkes [68], the acid–base properties of solid surfaces can be measured by the calculation of the work of adhesion against model acidic or basic liquids. In the second approach, described by Good and van Oss [69], a main parameter characteristic for each solid is calculated from measurements of wetting and contact angle. Equation (28) then allows for the determination of the surface energy components of the solid by measuring the contact angle of the solid substrate with liquids of known surface-tension components. Data reveal that Good and van Oss' approach to the evaluation of the $\gamma+$ and $\gamma-$ components of the solid surface free energy shows that most surfaces are overwhelmingly basic with a small acidic component, the so-called "catastrophe of basidity" [70]. In this connection, some researchers, e.g., Berg [71], suggest that to verify the method a more complete understanding of the wetting processes involving many interfaces and colloid phenomena is needed. Later, Volpe and Siboni [72] resolved this problem: they proposed a comparison of acidic components with acidic and basic with basic only, use of a wide set of solvents, and they emphasized that a complete and coherent definition of the calculation procedures is required.

Connor et al. [73] propose a criterion for optimum adhesion between two phases based on maximizing the wetting tension. It has been shown, that the maximum wetting tension is obtained when the substrate and adhesive surface energies are very high and equal. For example, in the acid–base approach, such a situation occurs when the Lifshitz – van der Waals components of the substrate and adhesive are equal and when the acidic component of one phase is equal to the basic component of the other phase. Comparison of these results and experimental data from other sources suggest that the wetting tension can be used as a criterion for optimum adhesion.

In adhesively bonded joints, there is a relation between the thermodynamic work of adhesion, W_A, and the total work of fracture, F. Recently, Penn and Defex [74] have determined W_A from the contact angle measurements made at room temperature with the adhesive in the liquid state and have established the relation between W_A and F. Values for F were determined from inverted blister tests conducted at temperatures low enough for the adhesive to be in the solid state. Because the thermodynamic work of adhesion, W_A, represents the energy required for reversible separation of the two materials at the interface, and because F includes the energy required for both reversible and irreversible processes during separation, the excess of F over W_A represents the energy consumed by the irreversible processes that occur in the specimen during loading and fracture. This excess, $(F - W_A)$, was found to increase exponentially with W_A and was attributed to the orientation hardening of the substrate within a very thin layer adjacent to the interfacial plane [74].

2.2.5. Weak boundary-layer model

The model of weak boundary layers was proposed by Bikerman [75], who claims that adhesive joints almost never fail exactly at the interfaces but at a weak boundary layer inside the adhesive or substrate. The author distinguishes several types of weak boundary layers: trapped air, contaminations at the interface, reaction products between adhesive components, or between components and medium. The theory proposes that clean surfaces can give sufficiently strong bonds with adhesive, but some contaminants such as greases and oils result in a layer formation that is cohesively weak. Moreover, it has been determined that certain small molecules and molecular fragments can migrate through polymers to reach the joined surface where severe deterioration of adhesive bond strength can result, sometimes reduced by a factor of ten.

The availability of the solid surface results in a decrease of the molecular mobility in the boundary layer as a result of the conformation-set-limiting and adsorption interactions of the polymer molecules with a solid at the boundary. This fact has also been confirmed by other researchers [76, 77] for filled epoxy polymers. It has been shown that geometric limitation of the number of possible macromolecules conformations close to the surface of the particles plays the fundamental role in the change of molecular mobility. As described by Duprée [58], two factors limit the molecular mobility of the chains close to the boundary: the adsorption reaction of the macromolecules with the surface and the decrease in their entropy. Close to the interface, the macromolecule cannot adopt the same conformation number as in the bulk, so that the surface limits the molecular geometry. Hence, the number of states available to the molecule in the surface layer decreases. Therefore, limitations of the molecular conformation are the basic reason for the molecular mobility decreasing close to the interface.

Lee [31] draws attention to Bikerman's assertion [75] that in any case of low strength with apparent residue-free separation, weak boundary layers are the main reason. Although it is asserted [78] that a weak boundary layer is a sufficient condition for low strength, it is not a necessary condition. In spite of Bikerman's model being criticized by several authors, e.g., by Lee [31] in the past, it is now admitted [46, 79] that many cases of poor adhesion can be attributed to a weak interfacial layer.

Recently, Yang et al. [46] successfully applied both the weak boundary layer and mechanical inter-locking models to the study of composite structures with weak-bonding defects by testing adhesively bonded, double lap-shear, and tensile specimens. The researchers show that a vibration damping and frequency measurement is an effective instrument for nondestructive detection of damage or degradation in adhesively bonded joints of composite structures.

Kalnins and Ozolins [79] study structure and some other characteristics of the boundary layer of a polyolefin adhesive, which forms the adhesion bond with steel under conditions of contact thermo-oxidation. They observe that the adhesion interaction of a polyolefin melt with steel leads to the formation of a cohesively weak polymer boundary layer. The structure of the boundary layer is less organized in comparison with the bulk structure, a fact that agrees well with the conception of a weak boundary layer offered by Bikerman [75].

2.2.6 Need for further study of adhesion mechanisms

During the last several decades much progress has been made in the study of the adhesion mechanism and the role of the interfacial properties and surface in the adhesion process and the mechanisms make and maintain adhesion. A series of theories have been developed to explain the process of forces in adhesively bonded structures. Separately, each of these theories is inadequate to completely describe the process of bonding in most situations. Each alone does not provide comprehensive global satisfaction. However, each theory contributes to an understanding of the overall mechanisms of adhesively bonded joint formation and, therefore, is important, especially for compound polymer–metal structures. While many interesting and important investigations have been done, there is still much to do to achieve a better understanding of this complex and multi-dimensional phenomenon.

2.3 Ultrasonic pulse–echo technique for adhesive bond-joint evaluation

The pulse–echo technique is one of the most widely used methods for inspection of adhesively bonded joints. A wide survey describing the "state of the art" in this field is given by Adams and Drinkwater [9]. Usually, pulses of longitudinally propagated compression waves or transversely propagated shear waves, with frequencies in the 1 MHz to 20 MHz range, are excited by a piezoelectric transducer. Ultrasonic pulses with these frequencies travel in the material being examined at wavelengths that provide a resolution of about 2.0 mm, at the lower frequency, to as high as 0.12 mm at the higher frequency, as shown in table 2.1 for the 20 MHz wavelength in the material. This short wavelength enables the detection of macroscopic anomalies, but is not short enough to achieve sufficient resolution for revealing microscopic anomalies.

Table 2.1 Selected Acoustic Properties			
Material	Longitudinal Velocity m/s	Wave Length at 20 MHz mm	**Z, Impedance** Kg/m²s X 10⁶
Aluminum	6320	0.316	**17.06**
Steel, 1020	5890	0.295	**45.63**
Adhesive in this Study	2290	0.115	**2.5 to 3.0** estimated
Glycerin	1920	0.096	**2.42**
Water	1480	0.074	1.48
Air	330	no transmission	**0.00043**

The wave train interacts with the material through which it travels, such that the pulse is modified by the material and any anomalies in the path taken. Thus the energy is impeded and/or reflected by discontinuities such as the adhesive layer, or the defects within it that were discussed in chapter 1, because these have a difference in acoustic impendence. Figure 2-6 is a schematic illustration of the longitudinal cross-section of a bond joint showing an adhesive layer containing several such defects. Note that the void and porosity defects located at B are not likely to reflect an echo, because the pulse is scattered by them.

The change in the acoustic impedance of the materials across the interface causes part of the wave to be reflected. Generally, defects contain air or other low-density substances with very low acoustic impedance relative to the adhesive. Table 2.1 lists the acoustic impedance for each material involved in this study. These values are the product of velocity and density.

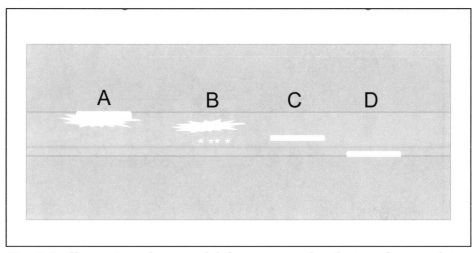

Fig. 2-6. Illustration of types of defects expected at the interfaces and in an adhesive layer. Note that defects A, C and D would reflect echoes,, but B would not, because the rough surface would tend to scatter the beam in many directions with little energy returning to the transducer.

Under common ultrasonic pulse-echo NDE inspection conditions, the ultrasonic signal will be almost completely reflected from unbonds, voids or other anomalies caused by missing material, because such pulse-echo inspections are usually performed with the transducer positioned normal to the surface, such that the direction of the ultrasonic pulse generated by the transducer is incident perpendicular to the surface of the specimen, as illustrated in the transducer arrangement shown in Fig. 2-1. This perpendicular arrangement simplifies the equation for calculating the reflection coefficient from that shown in Fig. 2-7, to the one shown in Fig. 2-8, because of the reduction of the angle of incidence to zero. The reduction of the angle of incidence to zero also reduces the magnitude of mode conversion concomitant with the angular entry of the beam at the interface between materials 1 and 2 in solids, where transversely propagated shear waves would be both transmitted and reflected.

By scanning the specimen and thus interrogating the adhesive layer and its boundaries at interfaces 1 and 2, information can be obtained about the area of adhesively bonded joints in the ultrasonic beam as well as various types of defects and morphological peculiarities.

A broadband transducer generates a sharp acoustic pulse, which propagates through the structure, and then collects the train of pulses reflecting from each interface. Such displays are commonly called A-scans. An example of an A-scan from a bonded specimen is shown in Fig. 2-9. The data available from the A-scan include the times of flight and amplitudes of various reflections, which may be individually gated and analyzed. The time of arrival of the echoes from the pulses are shown in the oscillogram of Fig. 2-9. The measurement of the time delay between two successive echoes provides time data to calculate the thickness of each reflecting layer of the specimen, when the velocity is known.

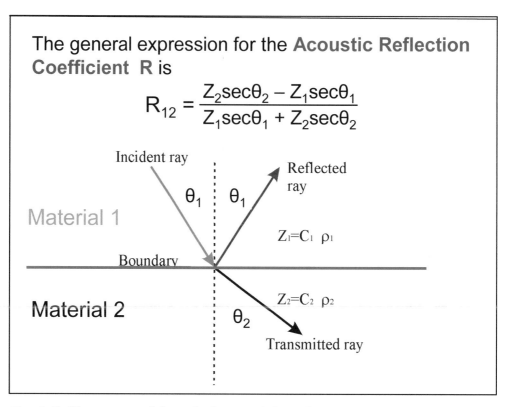

The general expression for the **Acoustic Reflection Coefficient R** is

$$R_{12} = \frac{Z_2 \sec\theta_2 - Z_1 \sec\theta_1}{Z_1 \sec\theta_1 + Z_2 \sec\theta_2}$$

Incident ray

Reflected ray

θ_1 θ_1

Material 1

$Z_1 = C_1\ \rho_1$

Boundary

Material 2

$Z_2 = C_2\ \rho_2$

θ_2

Transmitted ray

Fig. 2-7. Illustration of the calculation of the reflection coefficient at an interface.

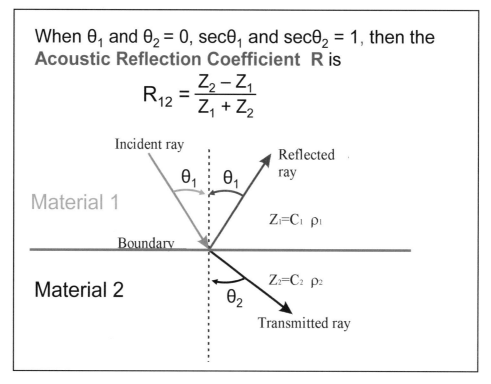

When θ_1 and $\theta_2 = 0$, $\sec\theta_1$ and $\sec\theta_2 = 1$, then the **Acoustic Reflection Coefficient R** is

$$R_{12} = \frac{Z_2 - Z_1}{Z_1 + Z_2}$$

Incident ray

Reflected ray

θ_1 θ_1

Material 1

$Z_1 = C_1\ \rho_1$

Boundary

Material 2

$Z_2 = C_2\ \rho_2$

θ_2

Transmitted ray

Fig. 2-8. Illustration of the calculation of the reflection coefficient at an interface, when the signal enters normal to the interface or boundary.

The correlation between acoustic and mechanical properties is described by the following equations [80–82]

$$Z = C\rho, \qquad C_L = \left(\frac{K + \frac{4}{3}G}{\rho} \right)^{\frac{1}{2}}, \qquad Cs = \left(\frac{G}{\rho} \right)^{\frac{1}{2}} \qquad (2\text{-}30)$$

where

Z is the acoustic impedance for the specific material medium,

ρ is the density of the medium,

C is the velocity of sound,

C_L is the longitudinal sound velocity,

C_S is the shear sound velocity,

G is the shear modulus, and

K is the bulk modulus.

Values for Z are shown for selected materials in table 2.1, along with values for C_L and the corresponding wavelengths for a frequency of 20 MHz propagating in the material. Note that Z is the product of a vector and a scalar; therefore Z, and the values of R computed from it, are vectors that will determine phase relationships in reflected echoes to be modeled and discussed in chapter 4, as well as observed and reported in chapter 5, when reflections from unbonded and bonded interfaces are encountered. Moreover, the relationship between acoustic velocity and modulus, as expressed in (30), can be valuable in predicting the mechanical properties of a bonded assembly.

A single A-scan contains quantitative information about the layer thickness, or when thickness is known, the transit time allows the sound velocity to be calculated. In previous local investigations [83], however, and work by other researchers show that the complex structure of the reflected waves makes the interpretation of individual A-scans difficult. At the same time, collecting a set of A-scans often makes it clearer which pulse is carrying the useful information.

The basic data may be individually gated and analyzed. By using the transducer-specimen arrangement shown in Fig. 2-1 to acquire ultrasonic pulse–echo data, such as that shown in Fig. 2-9, testing the quality of adhesive bonding for missing adhesive, uncured adhesive, adhesive partly cured in a pre-jelling oven, burnt adhesive near spot welds, and bonds modified to simulate cohesion and adhesion failure can be accomplished. The transducer, at normal incidence, serves as a transmitter–receiver, i.e., sound from the contact transducer passes through the sample, is reflected from the interfaces and subsequently detected by the same transducer.

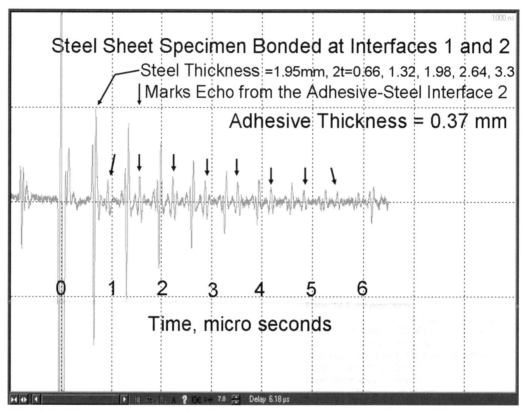

*Fig. 2-9. Ultrasonic pulse-echo A-scan of an adhesively bonded steel lap joint, with positive-phase echoes from the **bonded** adhesive-steel interface 2 identified by arrows. Values for 2t, the round-trip time of flight are shown near the top in μs.*

A promising method for the ultrasonic testing of adhesively bonded joints with A-scans, in the frequency range of 20 MHz–50 MHz, has been proposed by Goglio and Rossetto [84]. They used the acoustical methods to detect zones affected by poor adhesion. In these zones, the adhesive can be present in the gap between the sheets but bonding does not take place due to insufficient curing or non-wetting impurities on the interface. The reflected signal has the form of a damped succession of pulses as a result of the ultrasound reverberations inside the metal sheet. The decay, θ, of the signal is defined as the ratio of successive peak amplitudes [84], by

$$\theta = \frac{1}{n} \sum_{1}^{n} \left| \frac{A_{k+1}}{A_k} \right| = R_{MC} R_1 e^{-2ah} \frac{1}{n} \sum_{n}^{1} f(k) \qquad (2\text{-}31)$$

where

A_k and A_{k+1} are the amplitudes of the kth and $(k+1)$th peaks,

n is the number of ratios considered,

R_{MC} is the reflection coefficient between the sheet metal and coupling medium,

R_1 is the reflection coefficient at the first interface, (metal-air, R_{Ma}, or metal-adhesive, R_{MA},

α is the attenuation coefficient of the metal,

h is the thickness of the sheet, and

$f(k)$ is a function of the peak order k.

If $R_1 = R_{Ma}$ (the reflection coefficient for a metal–air interface), when adhesive is not present.

If $R_1 = R_{MA}$ (the reflection coefficient for a metal–adhesive interface) good adhesion exists.

The method allows an adhesion index to be evaluated that is related to the reflection coefficient at the adhesive–substrate interface. Statistical distributions of such indices can then be used to develop guidelines for evaluating the quality of adhesion in a sample [84].

In a subsequent work, Goglio and Rossetto [85] investigated bonded joints with anaerobic adhesives where the thickness of the adhesive layer is nominally zero due to contact pressure between the sheets. In these systems, the ultrasonic signals traveling through the interface are not completely damped. In such cases, the method based on the decay measurement described above is not suitable because the adhesive layer is negligible or absent. To overcome this problem, the authors take the adhesion index to be the ratio of the first peaks reflected from the bonded area after and before bonding. Using this so-called "first-peak" method allows the authors to extend the limits of the pulse–echo technique application to ultrasonic inspection of adhesive joints between sheets of thickness more then 1 mm, bonded together by an adhesive layer of nominally zero thickness.

Challis et al. [86] applied a transfer-matrix approach based on the Thompson [87] and Haskell [88] formulation for a bonded joint excited by a compression wave probe with a coupling shoe attached. The authors incorporate complex wave numbers in the layers to include the frequency-dependent effects of absorption and phase velocity in the adhesive. One of the difficulties in using the pulse–echo method for the investigation of adhesive joints is that the incident signal overlaps and interferes with the reflected signals, thereby making analysis of the received pulse sufficiently difficult. To improve the detectability of the low-amplitude echo from the rear adhesive–substrate boundary, it was necessary to reduce the amplitude of the reverberation from within the front metal sheet. Challis and et al. [86] have approximated the behavior of the echo and reverberations by a simple reverberator expressed in the frequency domain as

$$H(\omega) = \frac{1}{1 - \alpha e^{-jw2T}} \qquad (2\text{-}32)$$

where

α is the product of the internal reflection coefficient on either side of the front substrate boundaries and

T is the propagation time for the compression wave to cross the area of the sheet.

Because the signals are digitized, (32) can be represented in the z-transform domain, commonly used in digital signal processing [89], as

$$H(\omega) = \frac{1}{1 - \alpha z^{-2m}} \qquad (2\text{-}33)$$

where

m is the number of digitizer time steps equivalent to T.

The inverse of this filter has the following time-domain equivalent

$$y_n = x_n - \alpha x_{n-2m} \qquad (2\text{-}34)$$

where x_n and y_n are the input and output data, respectively.

Equation (34) describes a digital filter that clears away reverberation oscillations and clarifies the output signal from the lower interface. After filtering the data, the signal should only contain the bond line echoes, which can be processed using conventional pulse location and gating methods. Experiments, which have employed this model with 0.7 mm epoxy – 1.2 mm steel joint, have shown that front disbonds are testable in most cases. However, detection of a disbond from a rear substrate void is also possible after significant signal processing and for restricted combinations of substrate and adhesive thickness. Evaluation of the state of the adhesive cure was uncertain due to large errors in the calculations of the adhesive impedance in the measurements. Clearly, the results of these investigations are highly relevant for the automotive industry, where the nondestructive evaluation of simple substrate metal lap joint is commonly necessary.

For the inspection of adhesive joints consisting of an aluminum alloy substrate and epoxy adhesive, Vine et al. [90] used ultrasonic measurements at normal and oblique incidence. The oblique incidence scans were applied for detecting the same types of defect as the normal incidence technique, except in cases where the resolution was insufficient to detect very small defects. In the case of oblique incidence scans, no defect was detected that could not be observed by means of the normal incidence technique. In comparison to the normal incidence method, the oblique incidence testing tends to be more sensitive to the interlayer characteristics and more complicated to interpret because mode conversion can occur at the various interfaces. The authors performed ultrasonic scans, using a center frequency of about 50 MHz for normal incidence and 20 MHz for oblique incidence, on both two-layer and three-layer adhesive bond-joint specimens. In the two-layer specimens, micro-defects and edge disbonds were detected. The edge disbonds were easily found with both methods. The micro-defects were detected in areas remote from the edges and these isolated defects took several forms. The normal incidence scans allowed for the detection of many of these micro-defects, but only the largest of them could be observed using the oblique incidence technique. This was attributed to the poorer spatial resolution concomitant with the oblique incidence technique. It was found that in the three-layer compositions of 4 mm aluminum – 0.2 mm epoxy – 4 mm aluminum, no micro-defects were detected by any ultrasonic scans. At the same time, it is possible to assert that micro-defects can go undetected because of insufficient spatial resolution of the ultrasonic technique [91].

An approach that uses ultrasonic normal (7–22 MHz) and oblique (7–28 MHz) incidence techniques was applied by Moidu et al. at the University of Toronto [91] for the nondestructive durability study of epoxy adhesives using reflection measurements from the interfacial region between the adhesive (thickness at 0.6 mm) and the substrate (1.6 mm aluminum alloy). They used the spring boundary conditions [92] to model the interfacial area with tangential and normal spring constants. As a rule, the spring model has been used in cases where the adhesive and substrate may be treated as connected by normal and tangential linear spring elements [93]. Material nonlinearities will be discussed in the following chapter.

Moidu et al. [91] applied the above-mentioned approach for adhesive–substrate systems and showed that shear waves are a more sensitive indicator of the durability of epoxy adhesives than measurements of longitudinal waves with a given frequency at normal incidence. The shear waves at oblique incidence also showed somewhat greater sensitivity to degradation in comparison to shear waves at normal incidence.

In a comprehensive survey of nondestructive testing methods for inspection of adhesively bonded joints, Munns and Georgiou [11] note that a technique that uses shear stresses is likely to be more sensitive to interfacial properties than a technique that uses compressive stresses. An oblique incidence method commonly works as a practical inspection technique in cases where a high probe frequency is not required. Clark and Hart [94] have shown that for normal incidence shear waves are more sensitive then longitudinal waves for detecting small liquid-filled gaps in layered systems. Thus, the oblique incidence method can potentially detect the more gross types of defects, such as voids, cracks, and disbonds.

2.4 Resonant ultrasound spectroscopy

The method of resonance ultrasound spectroscopy (RUS) is well known and widely used for inspection of adhesively bonded structures. RUS is the technique of measuring the amplitudes of the natural ultrasonic frequencies of vibration (normal modes) of a sample. This spectral information can then be used to determine the material properties of the sample by the use of an appropriate model. Excitation can be achieved by a piezoelectric or eddy current transducer that is driven by either swept frequency AC voltage or a short pulse that overlaps the full frequency region of interest because of to its broadband nature. Either the same or a separate transducer can be used to detect the sample's response. After amplification the output signal $f(t)$ can be converted to spectra $u(\omega)$ directly by the spectrum analyzer or by using fast Fourier transform (FFT)

$$u(\omega) = \int\limits_{-\infty}^{\infty} f(t)e^{iwt}dt \qquad\qquad (2\text{-}35)$$

The spectra that are obtained usually contain a set of peaks, each of which corresponds to an acoustical resonance. Typically, smaller objects (millimeter-size) have resonance frequencies in the megahertz range, while larger objects (meter-sized) have the lowest resonance frequencies, as low as a few hundred hertz [95]. The frequency of the peaks is determined by the size and shape of the sample and its elastic properties. In this way, acoustic parameters can be calculated directly for uniform samples with a given shape and material parameters or, inversely, the geometrical and material data for known materials can be found from acoustical measurements. This latter fact makes RUS a good instrument for nondestructive evaluation purposes. At the same time, the oscillation patterns that occur in samples with complex structure often have a complicated character. The successful interpretation of the data obtained in this case demands a deep understanding of nature of the resonance process in a particular sample. Because a proper inversion algorithm for each particular sample may be required, the widespread use of RUS is somewhat limited.

The realization of the RUS method has been especially successful for material characterization tasks. Many researchers have measured the sound attenuation and elastic constants of crystals and solid materials, e.g., geological samples [96, 97] and stresses or metal fatigue in metal structures [95, 97]. Ultrasonic spectroscopy has also been used in chemical weapon verification and treaty compliance monitoring [95]. The basic principles and requirements for an RUS system are brought together in number of reviews [96, 98, 99]. The measurement of the lowest 50 resonances of a single crystal can be enough to precisely calculate all independent elastic moduli of the specimen [98]. Impulse excitation followed by a Fourier transform has a few disadvantages in comparison with swept CW excitation. The impulse method has low power per unit bandwidth due to spreading pulse energy over the full frequency range, a low duty cycle due to short pulse duration, and a relatively poor signal-to-noise ratio as a result of the broadband nature of the receiver. Also, the whole measurement system cannot have any resonances in the frequency range of interest, so it demands a special choice of undamped, metallic diffusion-bonded transducers and suspension that is only weakly coupled to the specimen. Practically all of these restrictions are valid for the examination of the quality of adhesive joints by RUS. But due to anisotropic nature of the adhesive and complex layered structure of an adhesive joint, a special approach is required in addition to dealing with the significant size and complex shape of the spectra of typical adhesive joints.

The data collected in the RUS method are the amplitude and frequency of a particular resonance peak. The factors that determine the sensitivity of this technique include the central frequency and bandwidth of the excitation pulse relative to the resonance we need to analyze, the distinct separation of the resonance of interest from nearby resonances in the sample, and the existence of a distinguishable difference between the amplitude and resonance frequency between the "good" and the "defective" part. Although it is less clear what factors limit the resolution of the technique, it is fair to say that the scanning system and beam width must play an important role, as they did in the pulse-echo imaging method. Accuracy of the method is determined by the instrumentation, coupling of transducers to the sample, accuracy of the sound velocity measurement (its precision should be higher than the expected resonance measurement), and the quality of the surfaces of the plate.

The transducers used in the RUS technique should also lack resonances in the region of interest, and contribute minimal noise to the measurement [99]. Both characteristics are achievable by using undamped transducers constructed with high-sound-speed materials. Sometimes damping of the probe is necessary over a relatively narrow frequency range because its bandwidth must overlap the whole frequency range of interest with minimum variation of sensitivity [86]. Instruments working on the RUS principle typically use transducers operating in the frequency range 0.1–10 MHz. Transducers that have a wider frequency range can monitor more resonance modes and, therefore, can extract more information about bond joint from the raw data.

In contrast to others ultrasonic techniques, very good coupling between transducer and specimen is not advantageous because it tends to shift the resonance frequency [99, 100]. On the other hand, very poor coupling reduces sensitivity. Hence, a compromise must be found by construction of a special probe with a protective layer specifically designed for RUS.

The general theory of RUS derives the amplitudes and frequencies of resonance from the mechanical Lagrangian of an elastic solid [101]. The adhesive specimens usually have a specific shape and can be approximated by a layered plate. The sets of resonances for the plates can be divided into two parts: low-frequency plate modes and relatively high-frequency thickness modes. The plate modes are primarily bending vibrations that are weakly dependent on adhesion properties. Despite this fact, a number of publications [102–104] have reported tests using the so-called membrane resonance method. This method works for the relatively large area of the adherend sheet. The layer above the defect can be modeled approximately as a membrane with thickness h clamped on edges and free in the middle. The first resonance frequency f of such a round disc with radius a is [102]

$$f = \frac{0.47h}{a^2} \sqrt{\frac{E}{\rho(1-v^2)}}$$

(2-36)

where

E is Young's modulus,

v is Poisson's ratio, and

ρ is density.

Equation (36) permits the direct calculation of the size of the voids for known thickness and material parameters. The weak acoustical link of these areas with the rest of the surrounding structure gives rise to the effective excitation of these low-frequency vibrations. The increased response from membrane-like zones is typically greater than a factor of 10 (20 dB). A number of practical devices have been created on the basis of this principle. The maximum operating frequency of these instruments is typically around 20–30 kHz, and this restriction limits the size of the smallest detectable delamination.

Resonance spectroscopy is a very powerful tool for the nondestructive testing of plates or sheets. Through-thickness vibrations are easily explained using the wavelength approach [102]. If the sound velocity of the material is known, resonance-frequency data allow the thickness of the plate to be calculated. If the fundamental resonance frequency of the plate with thickness t is f_0 with corresponding wavelength λ_0, then the thickness, h can be expressed as [101]

$$h = \frac{\lambda_0}{2} = \frac{c}{2f_0} \qquad \text{or} \qquad h = \frac{c}{2\Delta f} \qquad \qquad (2\text{-}37)$$

where

$f = f_{n+1} - f_n$, the difference between the *(n +1)*th and *n*th harmonics.

The set of the harmonics is equally spaced for a solid plate.

An adhesively bonded joint represents several plates glued together, so it can be considered as multilayered system. The acoustic response from this system is much more complex than for one single plate and the peaks are no longer equally spaced. The natural frequencies of the system depend on the material properties (density, elastic properties of the material), the geometry of the specimen, the thickness of the adhesive and adherend layers, and the boundary conditions. Hence, changing these conditions will affect the amplitude, width, and position of the peaks as well as resonance response as a whole. Detecting adhesive joint properties by the RUS method is based on the idea that the resonance response from the disbond area is equal to the response from a single thickness of the plate whereas the perfect bond spectrum will show additional peaks corresponding to double thickness vibrations. Intermediate states of bonding show intermediates spectra. A simple mathematical model for this behavior is described by Brown [97], where vibration characteristics are calculated on the basis of a mass-compliance-mass device.

The critical point in the adhesive joint is the presence of an interlayer with thickness of the order of 1 μm between the bulk adherend and the bulk adhesive. Its properties play a dominant role in the determination of adhesion strength and durability. The RUS method has been successfully used for the evaluation of the interface properties of adhesive joints.

One important method for characterizing thin interface layers and partially contacting interfaces is the measurement of the reflection coefficient. This measurement is mostly done in the time domain where amplitude of the reflected signal is monitored. However, more information is available in the frequency domain [23]. The reflected signals from a reference interface and the test interface were captured and processed in the following stages.

- FFTs were calculated for both signals to obtain frequency spectra.
- The measured spectrum was divided by the reference spectrum to obtain a normalized spectrum.
- The normalized spectrum from the glass–adhesive interface was divided by the normalized spectrum from the glass–air interface. This step produces a reflection coefficient spectrum.

Drinkwater and Cawley showed [23] that the reflection coefficient decreases during epoxy curing and reaches an asymptote that indicates the final reflection coefficient, which is significantly higher for the joints with defective bonds. They described four cases in which reflection coefficient measurements were made from an aluminum–aluminum interface and an adhesively bonded interface.

Cawley [105] compared two promising techniques for the adhesive interface quality evaluation (measurements of the reflection coefficient amplitude from the adherend/adhesive interface and measurements of the frequencies of the zeroes of the reflection coefficient from the adhesive layer) and showed that they both are sensitive to the changes in the interlayer between the metal and adhesive. The reflection coefficient zeroes are simpler and more reliable than reflection coefficient amplitude measurements. But at the same time, it is more sensitive to the variations in the bulk cohesive properties than the interlayer properties, thereby making the extraction of the properties of the interlayer by itself problematic. The experimental measurement of the amplitude of the reflection coefficient allowed an oxide interlayer in adhesive joints to be detected [106]. The best measurement was achieved in the longitudinal–longitudinal reflection coefficient at normal incidence and at 55° in the adherend, and shear–shear reflection coefficient at 32° in the adherend. The sensitivity of the measurement strongly depends on the oxide-layer porosity.

Wang and Rokhlin [107] also used the ultrasonic angle-beam technique instead of normal incidence for the evaluation of the interfacial properties of adhesive joints. The frequency response of obliquely incident ultrasonic signals from an adherend–adhesive interface was measured. Wang and Rokhlin developed a theoretical model for the analysis of the interaction between the obliquely incident ultrasonic waves and the multilayered adhesive joints. A special ultrasonic goniometer using only one ultrasonic transducer was built to measure the reflected signals. It focused obliquely-incident ultrasonic waves at different incident angles on the same area of the interface. The experimental results were in good agreement with those predicted from the model. Some of the minima in the reflected frequency spectra were independent of either the thickness or the elastic properties of the interfacial layer. This property enabled a simple and stable reconstruction procedure to be developed to determine thickness and interfacial properties from experimental data.

Adler, Rokhlin, and Baltazar [108] developed a scanning ultrasonic technique for the quantitative evaluation of adhesive bond joint integrity using Angle Beam Ultrasonic Spectroscopy (ABUS). This system uses, simultaneously, the oblique and normal incident beams on the bond line and measures the frequency response of the reflected signal. Bond thickness is determined by normal incidence data and thus thickness variations are accounted for by data analysis. Oblique waves are more sensitive to the interlayer properties and allow kissing or poor bonds to be discriminated from good ones. ABUS scans gives information about the inhomogeneity of the bond-line quality. A linear relation between the modulus and measured strength was also observed.

Yang et al. [109] developed a method for the assessment of adhesively bonded composite structures with weak joints by using vibration damping and frequency measurements. Damping is caused by the dissipation of the energy in dynamically loaded structures or materials. Internal damping is sensitive to the microstructure of the material, so any damage or defect inside the structure alters the damping properties of the system. Degraded adhesive joints were prepared with poor adherend surface preparation. The combination of resonant ultrasound spectroscopy and the half-power band method were used to measure the damping loss factor, η, where

$$\eta = \frac{\Delta f}{f_n} = \frac{\Delta f}{n f_0}$$
(2-38)

and

n is the mode number,

Δf is the half power band width of the nth mode resonance peak,

f_n is the resonance frequency of the corresponding vibrational mode, and

f_0 can be calculated as $v/2h$ where v is the velocity of sound, and

h is the thickness of the specimen.

Yang et al. showed that modal frequencies decrease and modal damping increases with an increased amount of unprepared surface fraction (better adhesion). Development of a dynamic numerical model with consideration of contact friction at defect interfaces allowed the authors to suggest that damping increases can be attributed to the contact friction at the crack interfaces. Moreover, this phenomenon was concluded to be a dominant energy dissipation mechanism in the composite "sandwich" specimen with cracks.

The resonant ultrasound spectroscopy method was used by Robinson et al. [110] for dry-coupling, low-frequency, ultrasonic wheel probes developed for the inspection of three-layered adhesive joints. The method is based on the principle that the partially bonded joint has a much higher mode-1 frequency because the mass of the system is reduced. To decide which peak is the mode-1 resonance of the joint, the FFT of the signal received from the joint was divided by that from a simple plane reflector. This method increases the clarity of the peak from ringing at the expense of the peak at the centre-frequency of the transducer. The authors claim that the use of low frequencies less than 500 kHz aids in the coupling of the transducer to rough surfaces. In addition, the much longer wavelengths produce a less directed beam, thereby reducing the sensitivity of the probe to accidental misalignment. This procedure improves the robustness of the measurement system and so aids its use in the assembly plant environment. Lower frequency also improves dry coupling performance with lower reflection coefficients providing better coupling.

It is known that characterization of the thin composite samples is particularly complicated due the very strong echo from the front surface of the composite plate, which masks the interference pattern caused by disbonds. Kašis and Svilanis [111] proposed a signal-processing algorithm using frequency domain imaging for efficient detection and characterization of delaminations in thin multilayered composites. For this purpose, the Fourier transform was performed not on the complete received signal burst but on the residual part that was obtained by subtracting the reference signal reflected by a flawless region from the pulse received at any arbitrary point. Only the differential signature of the area is analyzed in this case, and the negative effect of the strong frontal echo is reduced. Even after FFT resolution along the z-axis is lost, examinations at different frequencies are able to resolve delamination at the different depths. A simplified mathematical model based on a matrix approach was developed and experimentally verified. The authors propose that this technique is not limited to the case of three-layered composite plates.

Valdes and Soutis [112] studied delamination propagation in composite laminates by monitoring modal frequencies. In this work, the resonance frequencies were determined by correlating the maximum sensor output to the frequency scale while sweeping through the frequency range of interest, considering that the maximum amplitude of vibration occurs at resonance. The resonance spectrum of the sample was then compared with the baseline spectrum of the undamaged structure. The number, position, and shape of the peaks were analyzed. Data were arranged in a semi-three-dimensional plot showing amplitude, modal frequency, and damage area. It was shown that all peaks shifted continuously to lower frequency values with increasing size of damage. Changes of the modal frequencies gave a good indication of the degree of damage, as detected by the C-scans.

A few aspects of ultrasonic bond testing are given by Curtis [113]. The author discusses the relation between strength of joint and acoustical parameters. There is no direct nondestructive way to determine strength but some parameters relating to it can be measured. These parameters may include elastic modulus, adhesive thickness, spatial distribution of stress, distribution of flaw sizes, and detection of weak interfacial layers such as chemical contaminants or corrosion products. Changes in the elastic modulus or thickness can be detected by resonance methods in a straightforward manner.

Generally, the presence of an adhesive changes the frequency of the longitudinal resonance in comparison with a solid part containing no inclusions with weaker adhesive joints having greater frequency shift. This idea was realized in industrial devices for measuring adhesive bond strength, of which the Fokker Bond Tester II (late 1950s) [23, 9, 102, and 113] is perhaps the best known. This device can assess the frequency shift as well as amplitude changes in the first two modes of through-thickness vibrations. The device operates in the frequency range between 0.3 and 1.0 MHz. A series of comprehensive tests and practical experience showed that the instrument performed fairly well for metal-to-metal joints. Small voids and disbonds at different depths in the multilayer joint can be detected as well. However, the device was much less successful for honeycombs. The voids in the adhesive layer were detectable, but the device could not adequately find poor bonds due to uncured or poorly cured adhesion (cohesive properties of the adhesive). A further problem occurs with structures of continuously varying geometry, e.g., tapered panels. This variation produces frequency shift and amplitude changes similar to those of bond abnormalities.

Another idea is that the adhesion strength shows the correlation with difference in energy distribution across the spectrum in short-pulse excitation. However, only cursory examination has been made of this hypothesis as applied to the cohesive strength variation [113]. The difference in the strength of the joint was simulated by varying adhesive thickness. The damping of some modes increased while others decreased, but no general algorithm was given.

Acoustic spectroscopy is used to calculate the modulus and thickness of the adhesive layer. Cawley demonstrated [114] that changes of the adhesive modulus from its nominal value indicate modifications of the cohesive properties, e.g., a problem in the curing process. The presence of porosity in the adhesive also reduces the adhesive modulus. The author developed a technique that is able to measure the adhesive modulus with an accuracy of ±6%, a value which was reached by analyzing frequencies of more than two modes. A reduction by 5% of the adhesive modulus corresponds to a 7% reduction in adhesion strength according the model by Alers and a 10% reduction according to the model by Rokhlin, which was discussed in [94]. The correlation between adhesive modulus and joint strength also depends on others factors such as adhesive type, joint design, and type of loading. Therefore, the authors did not claim this method as quantitative.

Whitney and Green [115] used the RUS method to monitor the degree of cure of carbon-fiber-reinforced epoxy composite during curing in a noncontact method. The resonant spectrum of the plate was measured periodically as the temperature was raised to the cure temperature and during the subsequent cure and cool-down periods. Because composite curing changes the boundary condition, the resonance spectrum changes as the composite cures. Curing of the composite panel causes significant shifts in the frequency relative to a plate at the same temperature with uncured composite material. Differences in amplitude, frequency, and damping are related to the degree of cure of the composite.

Finally, the RUS approach was used to characterize plasma-sprayed coatings including partially yttrium-stabilized zirconium, intermetallic materials, and metal–polyester composites [116], employing a method that is more suitable for measurements performed with low plate-thickness/ultrasonic-wavelength ratios. The acoustic properties of the specimens were studied by measuring the spectrum of overlapping back-echoes reflected from the thin plate. The velocity and the attenuation coefficient of longitudinal waves were then deduced from the characteristics of the spectrum at the resonance frequencies of the plate. These ultrasonic parameters are good indicators of differences between the microstructures of coatings prepared with various plasma spraying conditions and may also be used to follow the evolution of the coatings after thermal treatment.

2.5 Acoustic microscopy for imaging adhesively bonded joints

The historical aspects of acoustic microscopy development have been discussed by Quate [117], while the physical principles and applications of acoustic microscopy have been described in detail by Briggs [20]. Acoustic microscopy has been used widely over the last 15 years for a variety of purposes. In their reviews [118, 119], O'Niell, Maev, Zheng and Solodov summarized results related to the development of mathematical models of the propagation of ultrasound in heterogeneous and anisotropic systems, including adhesively bonded joints.

The increased operating demands of adhesively bonded joints require inspection for thermal, chemical, and mechanical properties with high spatial resolution. Because the contrast of ultrasonic images is mostly determined by the contrast between viscoelastic characteristics of the materials under examination, these ultrasonic techniques continue be used to obtain quantitative information about mechanical material properties. Furthermore, the formation of the adhesively bonded joint involves a great number of factors [120, 121]. Included among these factors are the formation of different types of weak layers between the adhesive and the substrate, as well as the formation of often unavoidable internal stresses in the adhesive layer. Reliable evaluation of these factors depends on the resolution of the scanning acoustic microscope (SAM). It is well known that the spatial resolution of conventional methods of acoustic microscopy is limited by the wavelength of the imaging ultrasound. This spatial resolution, L, is given by the Raleigh criterion [122] as

$$L = \frac{0.61\lambda}{\sin\theta} \tag{2-39}$$

where

λ is the wavelength and

θ is the angle between lens-to-focal-plane rays formed by the lens center axis and aperture edge.

The actual experimental spatial resolution can only approach that of this theoretical limit. The spatial resolution of the scan also depends on other features of the scanning system, such as the beam width, which is limited by the transducer (or array element) size, curvature and wavelength. The depth resolution, d, depends primarily on the width of the incident pulse. Generally, higher frequencies increase resolution by decreasing d [122] at the expense of a signal that attenuates more during penetration.

$$d = \frac{2\lambda}{(1 - \cos\theta)}$$ (2-40)

where λ and θ are as defined previously in (2-39).

Acoustic waves transmitted through the adhesive–substrate interface may be used to examine these internal interfaces, because each interface between different materials reflects a fraction of the incident wave. The fraction is determined by the reflection coefficient, R, defined and discussed in chapter 3. The most common method of investigation uses an analysis of the reflected intensity of a focused acoustic wave. To accomplish this task, both the emitter and detector of ultrasound are combined on one probe head. When a one-dimensional scan is performed along a line on the surface and the echo amplitudes that return from interior features is plotted as a function of position, a cross-sectional view, or "acoustic slice", of the specimen is obtained and is called a B-scan. When a two-dimensional scan is performed, a planar view of the bonded joints, or C-scan, is produced. In this way, it is possible to determine the distribution of defects by scanning the surface of the structure with a transducer.

The scanning acoustic microscope typically uses a frequency range of 3 MHz – 2 GHz for focused acoustic waves, which limits the spatial resolution to around a few micrometers. An increase of the acoustic image resolution cannot always be achieved by an increase in frequency, because the attenuation of the ultrasonic wave is proportional to the square of this frequency. Hence, when a wave of sufficiently high frequency is used as a probe, the wave may not penetrate far enough inside the specimen for information about its interior to be obtained.

The main parts of the acoustic microscope are a transducer and an acoustic lens. The latter is coupled through an immersion medium to the surface of the sample. This coupling material helps to reduce the extremely high impedance mismatch between the material and air. While there exist some cases where the use of couplant can damage the sample (e.g., fiber webs and textiles), the use of an immersion medium is adequate for many types of static measurements including spot-weld testing, medical echography, and adhesively bonded plates.

Many parameters are involved in the choice of a coupling fluid. The velocity and attenuation of the acoustic waves in the fluid as well as the chemical reactivity of the fluid with the materials constituting the lens and specimen must all be taken into account. Theoretical and experimental investigations regarding the acoustic parameters of coupling fluids in acoustic microscopy are described by Cros et al. [123]. In this study, the effects of electrolytic solutions of LiOH, NaOH, and KOH in water were considered in order to increase the sound velocity. This approach is possible because the velocity of compression waves in liquids increases with the strength of the interaction between molecules. Because ion–water interactions tend to be strong, the velocity will increase as a function of the ion concentration, the polarizing capacity of the cations, and the polarizability of the anions.

Yamanaka et al. [124] used the SAM for investigations at low temperatures. They used a methanol couplant to design a SAM that is capable of operating between +30 C and −94 C. It was proven that at a low temperature (−30 C) the sensitivity of the method is much better than at room temperature. Further investigations have shown that resolutions as high as 15 nm can be obtained by SAM operating near absolute zero using 15.3 GHz ultrasound and liquid helium coupling fluid [125].

An alternative to liquid couplant is a solid coupling medium. Transducers of this type commonly use a soft rubber between the transducer and specimen, which conforms to the surface undulations of the test surface when in pressed contact. Billson and Hutchins [126] demonstrated the application of a static, solid-coupled, longitudinal-wave probe operating using a synthetic rubber-coupling medium. Drinkwater and Cawley [127, 128] described a similar probe operating at a frequency range of 3.8–7 MHz but composed of a different low-loss rubber. These techniques were capable of operating in pulse-echo mode and showed promise for use in generating C-scan data without the risk of test piece contamination. This approach has shifted development away from the use of low-frequency, undamped, through-transmission devices, which use a thin layer of highly attenuative solid coupling, towards higher frequency, highly damped, pulse-echo devices where the solid coupling is used as a delay line.

During recent years, enormous effort has been exerted to make the transducers sensitive enough to transmit ultrasound without using physical contact [129, 130], i.e. air coupling only. Blomme et al. [131] report studies on a wide diversity of materials of both low and high acoustical impedance including flaws in an aluminum plate, spot welds on metallic plates, and ultrasonic reflection from an epoxy plate with a copper layer in the frequency range of 0.75–2 MHz. One limitation of this technique is that the object must have a flat surface and parallel entrance and exit planes to accommodate the requirement that the measurements occur at normal sound incidence.

The choice of couplant depends on the required resolution and depth penetration, as well as the properties of the sample's surface. The survey of the acoustical experiments in this field for the last decade indicates that the main way to launch ultrasound into a sample is still to immerse both ultrasonic transducer and test material in a liquid (usually distilled water). However, this immersion technique is not suitable for some metals, wood, paper, or any other porous material, and in these cases air or a solid coupling may be used. For extremely high resolution, a coupling medium of low attenuation such as super-fluid liquid helium may be used, although at the cost of higher impedance mismatch and reduced depth penetration. Among limitations of the air coupling technique is an upper bound on the usable ultrasound frequency, (typically below 5 MHz) and consequently the relatively low resolution of this technique.

Acoustic imaging is well suited to the adhesive bond geometry. The amplitude of acoustic pulses reflected from each interface is very sensitive to the physical properties of the media on either side as well as elastic conditions at the boundary itself. The images, obtained using a commercial SAM, demonstrate the possibilities of the technique for detecting various types of defects and also the difficulties in dealing with the metal–polymer acoustic mismatch.

By using 50 MHz (normal incidence) and 20 MHz (oblique incidence) probes for an aluminum/epoxy interface, Vine et al. [88] obtained acoustical images that show that a very large number of line and spot micro-defects can be clearly detected, although a much smaller number are seen on the oblique incidence scan as compared to the normal incidence scan. The authors develop the potential of ultrasonic imaging application for the detection of environmental degradation in aluminum/epoxy adhesively bonded joints. Photographs taken of a specimen exposed to water at a temperature of 50 C for periods of up to 18 months show that defects as small as 0.5 mm in diameter could be detected at normal incidence, while the minimum detectable size at oblique incidence was around 2 mm.

Cross et al. [132] investigated the 12 μm polyethylene terephtalate (PET) films coated with a 40 nm thick aluminum layer by high-frequency (600 MHz) acoustic microscopy. Due to the superficial layer thickness of a few micrometers, acoustic attenuation was negligible, and images were obtained with the resolution of about 2–3 μm in focus. The authors observed the variations of the adhesion with the thermal treatment as a function of the number of unbonded areas. These areas have been characterized as losses of adhesion between the layer of aluminum and PET film. Certainly, the acoustic images of the samples show sufficiently rich contrast behavior; however, this technique can be used only for the detection of very thin bonded joints and is inapplicable to the thicker layers of the adhesive and substrate that are common in heavier industries.

The penetration of chemical agents into adhesively bonded structures is of particular interest for these investigations. The result of aggressive media exposure becomes apparent as a cause of structural modification of the adhesive layer. This modification practically always contributes to a loss of strength due to the onset of micro-cracks and micro-voids in the adhesive volume as well as on the interface with the adherend. The origin of these imperfections is the penetration of active substances through the interface and the resulting change of the adhesive structure on the interface and in the bulk of the joint. Testing with the pulse-echo technique has shown that micro-voids become detectable on C-scans, providing there is careful setup of the data acquisition gates.

This has been a brief review of some applications of the SAM for imaging and quantitative characterization of adhesive joints. In spite of the fact that acoustic imaging techniques are important tools in nondestructive testing and material evaluation, only a limited number of researchers have used them to investigate adhesive joints. This limited application may be explained by the fact that the presence of the finely layered structure complicates visualization, and these complications may only be attributed to adhesive/substrate thickness and the viscoelastic properties of the material under investigation. It is well known that some composite materials, such as glass-reinforced plastics, highly attenuate ultrasound, and, in such cases, the thickness of the material that can be inspected ultrasonically is limited [11]. For this reason, the use of a probe with a centre frequency of 30 MHz for the normal incidence longitudinal wave and a probe with a centre frequency of 17 MHz for the oblique incidence technique are typical for analysis of various engineering structures by this technique. It is also noteworthy to consider the difficult challenges expected to be confronted should this technology be offered as an approach to adhesive bond NDE in a manufacturing environment. Therefore further investigation into its effectiveness for the application intended for this research is deemed to be beyond the purview of this study.

2.6 Guided waves for the characterization of adhesive bond joints

Lamb waves are elastic waves in freely vibrating plates [133]. They typically have wavelengths that are on the order of the thickness of the plate, although an infinite number of modes can exist and can occur in symmetrical (S) or asymmetrical (A) varieties with respect to the centre line of the plate. In the small wavelength limit, Lamb waves effectively degenerate into Rayleigh (surface acoustic) waves, which propagate on each surface of the plate. Investigations with Lamb and leaky Lamb waves have been carried out since their discovery in areas ranging from seismology to nondestructive testing, acoustic microscopy, and acoustic sensors.

The use of ultrasonic Lamb waves to evaluate adhesive bond joints has become a cost-effective, valuable approach to assuring quality in adhesively bonded assemblies. Lamb waves are the particular modes of vibration that propagate along plate-like geometries where the thickness of the plate or sheet medium is less than the wavelength. The waves are excited by one transducer and received by another, after being propagated along the plate medium. Lamb waves provide a fundamentally different approach from the pulse–echo or resonance methods. This is because the physical nature of Lamb waves is different from the bulk waves used by the other approaches. The main changes are that Lamb waves propagate down the plate instead of through it, and employ much lower frequencies, so that the wavelength exceeds the thickness of the plate medium, consequently a separate transmitter and receiver is necessary, but it also means greater flexibility in terms of transducer technology. Lamb waves are also highly dispersive, meaning that different frequencies travel with different phase and group velocities in the medium.

The spatial resolution for Lamb-wave techniques depends mainly on the source–receiver distance and is significantly less, in principle, than can be achieved from bulk waves. On the other hand, because the mode of propagation is along the plate, it is possible to test large areas efficiently. However, in adhesive bond NDE applications, Lamb waves are sensitive to the geometry of the bonded system as well as the visco-elasticity of the bond joints in the component parts. These properties make them excellent tools for probing adhesively bonded joints, using both attenuation and velocity changes as NDE parameters.

In ultrasonic Lamb-wave inspections of adhesive bond joints, the waves can be introduced into the bond layers by mechanically coupled transducers (including air-borne ultrasound), electromagnetic perturbations, or laser impingement on the surface of the joint. In any case, the acoustic wave propagation is guided between two parallel surfaces of the test object, which in this case is the layered elastic media of the bond joint. Because acoustic perturbations are fundamentally more responsive to mechanical properties than thermal or X-ray interrogation of the bond joint, low-frequency ultrasonic inspection has been recognized widely as the desired inspection approach for mass-production manufacturing applications. Other approaches can and may be used to compliment and (or) confirm the result, but the acoustic approach has the highest potential for yielding reliable results. The low frequency also helps in avoiding the addition of a liquid couplant to the surface.

When a bonded assembly is submerged for inspection by Lamb waves, leaky Lamb waves result. Leaky Lamb-wave propagation is induced by oblique insonification where a pitch–catch ultrasonic transducer arrangement insonifies a plate-like solid immersed in fluid. The phenomenon is associated with the resonant excitation of plate waves that leak energy into the coupling fluid and interfere with the specular reflection of the incident waves. The destructive interference between the leaky waves and the specularly reflected waves modifies the reflected spectrum introducing a series of minima in the spectra of the reflected waves.

In the last few decades, evaluation of adhesively bonded joints and heterogeneous structures using Lamb waves has been the subject of study of many researchers concerned with inspection of the bond quality. Rokhlin and Marom [134], Bar-Cohen and Chimenti [135], Achenbach and co-workers [136, 137], Mal, et. al. [138], Pilarsky et al. [139], and Nagy and Adler [140] made important contributions to the acoustical methods developed for the inspection of adhesively bonded joints such as those in multilayered sandwich structures. It has been widely demonstrated that the velocity and wavelength of plate waves are sensitive to the mechanical properties and to the boundary conditions between the adhesive and substrate.

Singher [141] has proposed a quasi-static model for ultrasonic guided wave interaction with an imperfect interface and applied this approach for indirect estimation of the bond strength. For bond strength evaluation two parameters have been chosen: the velocity of the ultrasonic guided wave and the difference in the characteristic frequency between the two recorded signals. Singher has shown that a good correlation exists between the velocity of the guided waves and the bond strength. This method can be used to classify the interface interaction and provide a useful nondestructive tool.

To develop a model of an adhesive structure with distributed voids in the bulk, Vasudeva and Sudheer [142] identified the specific Lamb modes that can characterize different kinds of defects in layered plates. They show that the modes for which high stresses occur at the interface indicate the existence of defects such as voids and pores whereas the modes for which the displacements are high show the existence of harder inclusions. Cowin and Nunziato [143] attempt to identify the use of a linear elastic material with voids (LEMV) model in nondestructive evaluation of bi-laminates via stress analysis. The dispersion curves of the Lamb modes obtained in the frequency range 0–10 MHz, using estimated values for the parameters of the LEMV, are quite similar to the ones given by other researchers.

The experimental results obtained by Heller et al. [144] demonstrate the effectiveness of combining laser ultrasonic techniques with the two-dimensional Fourier transform for characterization of adhesive bond conditions. They show the possibility of experimentally measuring transient Lamb waves in two types of bonded aluminum plate specimens as well as investigating the influence of the adhesive material conditions on the dispersion curves of these specimens. The dispersion data are used for monitoring the changes in the bond condition as a function of age. These measurements are possible, in part, because of the use of high-fidelity, broadband, point source/point receivers and the non-contact nature of laser ultrasonics.

In summary, a wide range of research and industrial techniques have been developed to tap the potential of the Lamb-wave methods for detecting changes in adhesive and geometrical properties in the interface layer. Since Lamb-wave characteristics are strongly dependent on the thickness of material through which they propagate, the measurement of the Lamb-wave velocity may be used to indicate a change in bond line thickness. Lamb waves also lend themselves naturally to remote generation techniques such as laser- and air-coupled ultrasound; these two areas remain the subject of significant ongoing research [136, 140, 145].

2.7 Literature search summary and conclusions

Analysis of the different applications of the SAM for characterization of adhesive joints shows that there are two principal categories of evaluations. The first category includes the imaging characterization of surface, subsurface, and bulk material comprising components and structures. The second category consists of the quantitative characterization of laminated structures and parameters of adhesive layers by means of analyzing reflected pulses. The SAM method allows evaluation of both microstructure and mechanical properties in a single examination. This advantage is a result of the specific nature of the acoustic contrast formation. In acoustic microscopy, the contrast of raster images is based on the variations of the local physico-mechanical properties, as opposed to the light and electron microscopy or X-ray methods, where contrast depends on electromagnetic properties of dielectrics. Using frequencies from 3 MHz up to 100 MHz, acoustic microscopy allows lateral resolution from $300\,\mu$m to as small as $10\,\mu$m to be achieved, along with axial resolution from $300\,\mu$m to as small as $5\,\mu$m. These parameters are extremely important for revealing the boundaries of disbonds, various delaminations, and other micro-defects. In addition to higher resolution, acoustic microscopy provides information about defect location, and internal layer thickness in multilayer structures. As a result, it is possible to evaluate mechanical properties and establish a correlation between the internal structural patterns and the micromechanical properties.

Resonance ultrasound spectroscopy is a suitable and convenient technique for nondestructive measurement of the elastic modulus of materials and the quality of adhesively bonded joints. All the main adhesion defects (disbond, poor adhesion, cohesion failure) and adhesion thickness can be successfully monitored by RUS. The most problematic defect to evaluate is poor adhesion, which requires special signal processing in most cases. Several papers [104, 108, and 109] have outlined some significant advantages, as well as some difficulties, in applying spectral methods to the imaging of certain common types of adhesive bonds. These advantages in particular include greater sensitivity and better contrast for the case of thin metal sheets bonded with a polymer adhesive. The technique takes advantage of the acoustic resonance occurring with a different frequency in each layer, separating the response of the layers by looking in the frequency domain.

In the case of array and matrix transducer technologies, the major criterion for the imaging capability of the system generally depends on the penetration depth of the transducer and the system's resolution. The influence of the various parameters on the quality of the ultrasonic imaging, including various interface conditions, is an important issue for research and requires further study.

Acoustic methods have been widely used for the investigation of adhesively bonded structures, and common ultrasonic techniques used have been reviewed here. They include:

      (1) normal and oblique ultrasonic scans,

      (2) resonant ultrasonic spectroscopy, and

      (3) Lamb-wave methods.

These three are of interest because they are applicable to the nondestructive detection of voids, delaminations, porosity, cracks, or poor adhesion in bond joints, but perpendicular pulse-echo ultrasonic scans and Lamb-wave methods are of special interest in this investigation because they can be used to evaluate bond joints rapidly from one side.

The advantages offered by a 20 MHz pulse-echo bond NDE technique include sufficiently high resolution, inspection from only one side, a small transducer allowing access to confined regions, moderate attenuation in the joint material allowing sufficient penetration, and of course transportable as well as portable instrumentation. These attractive advantages contributed to directing the focus of the investigation toward the development of a 20 MHz pulse-echo technique with phase-shift sensitivity in the reflected echo that till provide for the identification of the state of adhesion at each of the two adhesive interfaces.

3. Theory of the Propagation and Reflection of Ultrasonic Pulses in Adhesive Bond Joints

The excitation, propagation and reflection of ultrasonic waves in solid media are thoroughly studied phenomena that have been researched and reported in literature since early in the twentieth century. In 1931 Mulhauser obtained a patent for using ultrasonic waves with two transducers to detect flaws in solids. In 1940, Firestone, and in 1945 Simons, developed pulsed ultrasonic testing using a pulse-echo technique [146]. Medical and industrial applications have followed and continue to expand as researchers throughout the industrialized world develop new non-invasive ultrasonic techniques. This investigation seeks to advance the state of the art by focusing on two of the many aspects of ultrasonic wave propagation in solids, in order to accomplish a higher resolution technique and user-friendly methodology for NDE of layered media. The first is the phase-sensitive pulse-echo analysis that clearly identifies which adhesive surface is not bonded to the adjoining substrate. The second is the Lamb wave approach to the interrogation of adhesive bond joints.

The ultrasonic pulse-echo method of bond NDE is accomplished by the acquisition and analysis of acoustic echoes that return from the ultrasonic interrogation of bond joints that have interfaces between layers with large acoustical impedance mismatch, as was illustrated in the cross-sectional view of a bond joint in Fig. 2-1. These reflected echoes result from a pulse excited by a piezoelectric transducer and introduced through a liquid couplant into the material under inspection. The ultrasonic pulse travels through the first layer of the material. Echoes from the first interface reverberate in the top metal layer of the joint structure, as shown schematically in Fig. 3-1. A fraction of the pulse is transmitted through each interface of the bond joint, as illustrated for a half bond in Fig. 3-2, and echoes are reflected and return from each interface, and are detected by the sending transducer. The fraction of the signal reflected is determined by the reflection coefficient, R, and the fraction transmitted is determined by the transmission coefficient, T, which is 1-R. The transducer is equipped with a plastic delay line to provide ample time for damping, after the excitation pulse is sent, so that it will have become quiet and thereby have enhanced sensitivity to detect the returning echoes.

The phase-sensitive pulse-echo analysis is very effective, and provides high resolution, when significant differences in acoustic impedances exist between the materials comprising the bond-joint layers of the substrate(s) and adhesive. The Lamb wave approach is efficient and effective when significant differences in acoustic impedances do not exist between the materials comprising the bond-joint layers, and high resolution is not needed, nor is it required to identify which adhesive surface is not bonded to the substrate. The theory supporting each of these two approaches to the utilization of

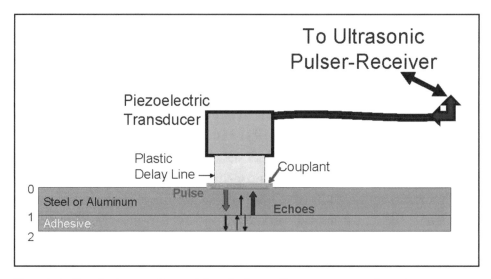

Fig. 3-1. *Single layer of metal bonded to adhesive at interface 1.*

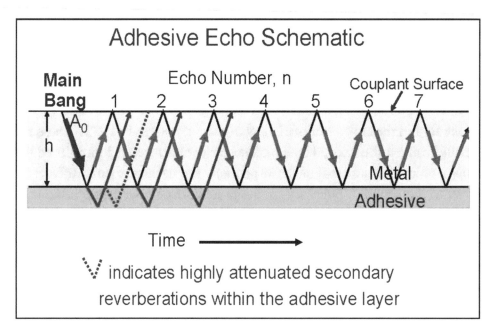

Fig. 3-2. *Reverberating echoes in a single layer of metal and adhesive,   bonded at interface 1. Reflected echoes return from metal-adhesive interface 1, the adhesive-air interface 2 and reverberate by reflecting from the metal-couplant interface 0, under the pulse-echo transducer.*

ultrasonic wave propagation and reflection to assess adhesive bond joints will be examined in this chapter, within the scope and constraints determined by the materials and geometric configurations intended in the application. These constraints include the normal orientation of the pulse-echo transducer on virtually flat surfaces of simple sandwich-type bond joints made by adhesively bonding steel, aluminum and polymer sheets.

A longitudinally vertical cross-section of a typical bond joint is schematically illustrated in Fig. 2-1. The arrows depict the pulse being introduced into the top-layer substrate by the transducer and traveling through the joint. Echoes return from each interface with a fraction of the pulse amplitude. A fraction of the pulse is transmitted and a complementary fraction is reflected at the first interface with the adhesive or the air, which ever is present there. The fraction of the pulse that travels through the adhesive is similarly transmitted and reflected at the adhesive-substrate interface, or adhesive-air interface, should no bond exist there. These reflected pulse-echo signals return to the surface, where the sending transducer detects them. The echo arrival times, amplitudes and phases are then analyzed to determine the bond state at each reflecting interface. The interaction of the ultrasonic pulses and the resulting echoes with the materials under interrogation must be understood in order to correctly interpret the echo characteristics, and draw from them the information needed for the characterization of the bond state at each interface. Hence, an examination of the physics that govern acoustic transmission and reflection will be undertaken here: first for the pulse-echo technique, then for the Lamb-wave approach. Each approach is based on the effect the properties of the bond joints have on the transmission and reflection of these ultrasonic waves.

3.1 Bond joint materials

The scope of the theoretical discussion will be limited to the materials and structures comprising the adhesive bonds to be evaluated in this investigation and in the ensuing applications derived from the methodologies developed; therefore the bond-joint materials that define those limits will be presented here.

3.1.1 Bond joint substrates

The bond joint substrates are engineering materials with suitable structural properties, such as modulus, strength and toughness, for cost-effective, durable vehicle construction. The materials included in this investigation are steel and aluminum sheet metals, fiberglass-reinforced composites such as thermoset polyester sheet-molding compound (SMC) and thermoplastic composites, as well as thermoplastics without reinforcements. These materials have experienced wide applications in vehicle body assemblies and with those applications come the challenge of assuring reliable bond joints.

3.1.2 Bond joint adhesives

A wide variety of bond joint adhesives and sealants are being used for applications that range from structural joining to sealing and noise reduction. These bonding and sealing materials tend to be more visco-elastic than the substrates. Moreover, in many cases the ultrasonic evaluation of the bond-joint integrity must be performed when the adhesive is in various states of cure that is before, during or after cure, when the visco-elasticity of the joining material may be even more pronounced. This wide window of variability of the material properties of the adhesives and sealants poses a challenge that must be addressed here in order to drive the development toward a successful implementation in each of a variety of applications. Therefore both the nonlinearity of wave propagation in the adhesive materials and the adhesive bond joints will be examined as a part of the discussion of theory. This is necessary because when high frequency ultrasonic waves are transmitted through materials, the ratio of the displacement amplitude to the wavelength, or period, is such that the half-cycle position conformity (full-amplitude excursions in 25 ns) of the transmitting elements of the material is impaired, thus giving rise to nonlinearity.

There are examples of nonlinearity in bond joints where the adhesive plays virtually no role. In diffusion bonds, for example, there is virtually no adhesive material. The bond is better regarded as simply an interface between two substrate media. Because there is so little material, little material nonlinearity should clearly be expected. It is reasonable, therefore, to look for different sources of nonlinearity. However, in this notable example it is observed that interfaces alone can be the sources of very significant, anomalously large nonlinear effects. The extreme example of this is the case of an unbonded, clapping interface [147, 148], which has been studied theoretically and experimentally for some time. The range of such nonlinear effects may be labeled as structural nonlinearities due to their relation to the performance of the structure itself as opposed to the material of which the structure is made.

Structural nonlinearities in adhesively bonded joints may arise in a number of ways. Common to all of these is the location of the structural defect, typically in the very thin layer of adhesive bonding, where particles of one material (adherend or substrate) are bound to particles of a second (adhesive) by adhesion. The overall strength of the joint, therefore, depends on the strength of these bonds. In some cases of poor adhesion, however, these bonds are weak or nonexistent, yet the close proximity of the two materials and the thinness of this layer mean that their influence on an acoustic wave is small. At the same time, it has been widely speculated that such defects should be highly nonlinear in that they might introduce significant amplitude-dependent distortions into any passing wave. A well-studied example of these cases is that of the zero volume disbonds or kissing unbonds. Although various practical definitions of these terms exist [149], as well as an extensive discussion of them in the introduction, the basic issue is that the two media are unbonded, while remaining in intimate contact, so that the interface supports a compressive but not a tensile load.

One common theoretical approach in dealing with this case is to use "unilateral boundary conditions" of one form or another [150–153]. Typically the kissing interface is treated as a pair of plane interfaces held in contact by some external pressure, with the probing beam being a plane harmonic wave. The normal welded or bilateral boundary conditions take the form of equalities enforcing continuity of displacement and traction across an interface. In contrast, unilateral boundary conditions typically involve conditions determined by inequalities governing changes of the state of the interface [149, 153]. If we are considering plane surfaces governed by a Coulomb friction model while in contact, the application of such boundary conditions results in predictions of a number of linear and nonlinear effects, including changes in reflection and transmission, phase, wave distortion, harmonic generation, and hysteretic energy loss, etc., dependent upon the applied pressures, wave amplitudes, material, and surface parameters. The situations described are essentially multi-linear, with an amplitude-controlled switching between linear sets of boundary conditions [154]. The next step to the realization of a more realistic interface model has been to add surface roughness to the mix.

Numerous models have been developed along these lines, both to give additional insight into closed disbonds in adhesive bonding problems [155, 156] as well as to predict nonlinear scattering from bulk cracks for nondestructive testing [157], and for seismic waves along fault lines [158, 159]. The main difference comes in situations where the wave amplitude is of the order of the surface roughness within the unbonded region. Nonlinearity within this regime is caused by the variation of surface contact and, therefore, elastic parameters with the pressure applied at the interface [156].

More recently, attention has turned to more phenomenological approaches specifically geared to the nonlinearity of the adhesive joint, possibly including, but not limited to kissing bonds. The idea is a heuristic one, trying to create a picture of the overall strength of the bond without resorting to details. Probably the oldest and best established of these is the so-called quasi-static approach (QSA), which was first described by Thompson and Baik [160]. The idea here has been to treat the interface as a set of (possibly nonlinear) springs tying one material to another. Often the inertial mass is ignored or set to zero, which may be reasonable if the adhesive material is light compared with the adherent material. In that case, the similarity of this set of boundary conditions with the unbonded interface (particularly the sliding or open cases) is more readily apparent. From this point, it becomes a question of the determination of these constants or functions, using either theoretical models [161–166], or by experiment [164–169].

Variations in adhesive properties can be acoustically detected by monitoring the linear properties of the acoustic signal: reflection and transmission coefficients, velocity or time of flight, and attenuation. The fundamental theoretical model supporting the interpretation of these properties is derived from the fact that the propagation of acoustic waves across linear layered materials is predictable, and has been well established in the literature for many decades now [156]. Variations can be found that seek to simplify the mathematics for particular cases, one of the most common approximations being a treatment of only "thin" adhesive layers [157–159]. In all, the theory regarding the detection of weak adhesion using linear means is well established, and a number of reviews have been written [11, 102, 119, 160]. It has been found over the years, however, that the sensitivity of these linear parameters to certain specific types of super-critical defects is not sufficient. These defects include such things as partial cures, closed or "kissing" unbonds, and generally weak adhesion.

Common to all of these is a low overall bonding strength combined with acoustical contact sufficient to transmit the low-power ultrasonic signals that are typically used in NDE applications. Because of the detection problems associated with these defects, a proper evaluation of adhesive joints using conventional ultrasonic techniques is inconclusive in many instances. For this reason, many researchers have sought to apply nonlinear acoustical techniques as an alternative approach or complementary methodology.

The discussion and analysis in the Introduction, however, of the unlikely probability of these "kissing" bonds, minimizes the importance and value of this aforementioned alternative approach for industrial vehicle manufacturing applications. Nevertheless, in these sandwich-type bonded structures, there are expected to be at least two possible sources of nonlinearity. The first source is the adhesive material itself, which in some cases might be an inherently nonlinear material, such as for example, a rubber-like adhesive. The second source might be structural nonlinearities in the adhesive bond line. These include weak bonds or "kissing" unbonds. While the material nonlinearity is not necessarily an indicator of the bond strength, it is an indication of the state of the material itself. The structural nonlinearity is, however, often thought of as being directly linked to the strength or weakness of the bond itself.

The effect of nonlinearity on acoustic wave propagation in materials has been investigated by several researchers. Work in this area by Humphrey[170] and other researchers concluded that in high amplitude ultrasonic fields, such as those used in medical diagnostic ultrasound, nonlinear propagation can result in waveform distortion, with concomitant phase shifts and the generation of harmonics of the initial frequency, thereby leading to complicated interpretation techniques. The issue is addressed here to raise awareness to it, but the degree to which nonlinearity is expected to impact the outcome in this investigation is minimal.

3.2 Propagation of ultrasonic longitudinal waves in adhesive bond joints

The simplest type of adhesive bond joint is a classical three-layered system where two, sometimes similar, substrate materials are joined by a layer of adhesive material between them. While the substrates are typically treated as materials having well-defined mechanical properties, the characteristics of the adhesive layer are much less so. The relative behavior of acoustic signals incident on and traversing this sandwich-type structure depends, therefore, on the mechanical (modulus, morphology, etc.) and geometrical (thickness, shape, etc.) properties of the substrate material(s) and the adhesive layer(s). It is therefore important to understand the transmission and reflection of ultrasound in both the substrate and adhesive layers in order to effectively utilize it as a reliable acoustic interrogation tool. Ultrasonic transmission in the bond-joint material is dependent on two important material properties: velocity and attenuation.

3.2.1 Velocity of acoustic wave propagation

The velocity of acoustic wave propagation in these adhesive bond joints must be considered, because it is a key factor in the correct analysis of echo signals from the substrate(s) and adjacent interfaces with the adhesive layers. The simplest expression for the velocity of a compression wave in an elastic medium is given by [171]

$$c_L = (B/\rho)^{1/2} \qquad (3-1)$$

where

c_L is the longitudinal velocity

B is the bulk modulus of the material and

ρ is its density

This simple, but inadequate, equation can be derived from the familiar linearized one-dimensional continuity equation, the linearized one-dimensional force equation and the equation of state, with conservation of mass and momentum to yield a second-order partial differential wave equation

$$\frac{\partial^2 p}{\partial x^2} - \frac{1}{c^2}\frac{\partial^2 p}{\partial t^2} = 0 \qquad (3-2)$$

in which the solution for p, the pressure wave front, is described by

$$p = p_0 \sin(\omega t \mp kx) \qquad (3-3)$$

and

$$c = \sqrt{\frac{B}{\rho_0}}. \qquad (3-1)$$

This one-dimensional equation for acoustic wave velocity is insufficient to explain the wave propagation velocity in the three-dimensional bond joints evaluated in this investigation. The failure of this equation is shown in the example for iron and its steel alloys. They have a body-centered cubic crystal structure, with density of about 7.87 Mg/m^3, bulk modulus of 140 GPa, shear modulus of 80 GPa and modulus of elasticity approaching 205 GPa [172]. Using the values supplied at this web site and equation (3-1), the value calculated for c_L in 1020 steel is 4218 m/s. Reliable sources, confirmed by experimental data reported in chapter 5, give a c_L value of 5890 m/s. Obviously, this simple equation does not yield correct results. This compels a closer examination of the relationship between c_L and the mechanical and physical properties of the material. This examination can begin by a look at the acoustic wave equation in three dimensions [173]. The second order partial differential equations describes the evolution of pressure, p, and velocity, u, as functions of space, r, and time, t, with the displacement vector, r, representing x, y, z. Hence,

$$p = p(\mathbf{r},t) = p(x,y,z,t) \tag{3-4}$$

$$\mathbf{u} = \mathbf{u}(\mathbf{r},t) = \mathbf{u}(x,y,z,t) . \tag{3-5}$$

Thus it is seen that the motion of the pressure wave, the function defined by p, and the motion of the particles, described by the function u, within the material are time dependant. The three-dimensional velocity equation in p is

$$\nabla^2 p - \frac{1}{c^2}\frac{\partial^2 p}{\partial t^2} = 0 \tag{3-6}$$

Variations in pressure, p, a scalar, and the displacement, u, a vector, are governed by wave functions. The displacement equations have been rigorously derived by Denisov [174] for both longitudinal and shear acoustic wave propagation, and can be written as

$$\frac{\partial^2 \boldsymbol{u}_l}{\partial t^2} - c_l^2 \nabla^2 \boldsymbol{u}_l = 0 \tag{3-7}$$

where

$$\nabla \times \boldsymbol{u}_l = 0 \tag{3-8}$$

and

$$\frac{\partial^2 \boldsymbol{u}_s}{\partial t^2} - c_s^2 \nabla^2 \boldsymbol{u}_s = 0. \tag{3-9}$$

where

$$\nabla \cdot \boldsymbol{u}_s = 0 \tag{3-10}$$

and arriving at

$$\rho\frac{\partial^2 \boldsymbol{u}_l}{\partial t^2} - (\lambda + 2\mu)\nabla^2 \boldsymbol{u}_l = 0, \tag{3-11}$$

and

$$\rho\frac{\partial^2 \boldsymbol{u}_s}{\partial t^2} - \mu\nabla^2 \boldsymbol{u}_s = 0. \tag{3-12}$$

The Lamė constants, λ and μ, are related to the more familiar elastic constants, Young's modulus E, shear modulus S, Poisson ratio ν, and bulk modulus B, by

$$E = \frac{\mu(3\lambda + 2\mu)}{\lambda + \mu} \tag{3-13}$$

$$S = \mu \tag{3-14}$$

$$\nu = \frac{\lambda(\lambda + \mu)}{2} \tag{3-15}$$

and

$$B = E/3(1\text{-}2\nu) \tag{3-16}$$

The acoustic velocities can be expressed in terms of the Lamė constants [175] by

$$c_L = ((\lambda + 2\mu)/\rho)^{1/2}, \tag{3-17}$$

$$c_s = (\mu/\rho)^{1/2} \tag{3-18}$$

and the ratio of the two velocities can be expressed in terms of Poisson's ratio, ν, only:

$$c_L/c_s = (2(1\text{-}\nu)/(1\text{-}2\nu))^{1/2} \tag{3-19}$$

Thus it is shown that the longitudinal wave velocity must also include a contribution from the shear modulus, as would be expected in a solid where waves are propagated by deformation stresses and concomitant strains that are not confined to one dimension. So the relationship between acoustic and mechanical properties of the material is, as was described by equations (2-30) and repeated here for convenience:

$$c_L = \left[\left(B + \frac{4}{3}S\right)/\rho\right]^{\frac{1}{2}}; \quad c_S = \left(S/\rho\right)^{\frac{1}{2}} \tag{3-20}$$

where

ρ is density

c_L is longitudinal acoustic velocity,

c_S is shear acoustic velocity

B is bulk modulus,

S is shear modulus,.

Using the previous values for bulk modulus and shear modulus in this equation yields a c_L of 5598 m/s in 1020 steel, within 5.2% of the 5890 m/s accepted value for 1020 steel.

Velocity can also be influenced by grain size. Precise ultrasonic velocity measurements were used to estimate average grain size in an AISI type 316 stainless steel by using the pulse-echo-overlap technique to measure ultrasonic transit time. The results indicate that grain size can be predicted with good confidence level using ultrasonic velocity measurements. Shear waves are found to be more sensitive for grain size measurement than longitudinal waves, and are said to yield grain size estimates by velocity measurements that are more accurate than those obtained by attenuation measurements [176].

3.2.2 Attenuation during acoustic pulse propagation

The grain size in metals are micro-structural features that significantly influence their mechanical properties, and hence their acoustic transmission characteristics. [177] Because most rolled and stamp-formed sheet metals have anisotropic elastic constants and neighboring grains have different orientations, with each grain offering slightly different acoustic impedance from its neighbor, ultrasonic waves are scattered during their transmission through these polycrystalline aggregate materials, thus resulting in attenuation.

According to the data shown in table 2.1, the prevailing wavelength of the 20 MHz ultrasonic signal in the metals used in this investigation, aluminum and 1020 steel, is 0.3 mm. The grain size measured in 1020 steel is 90 µm, with a standard deviation of 30 µm [178]. When the wavelength is more than three times greater than the grain sizes, as is the case in the 1020 steel, and the frequency is constant, the attenuation is confined to two of the three regimes available:

> (1) the Rayleigh regime where the wavelength is much greater than the grain size and the attenuation increases as its cube, and
> (2) the stochastic regime where the wavelength is about the same as the grain size and attenuation increases in proportion to it.

The third, the diffusion regime where the wavelength is much less than the grain size, is not included in the discussion here, because the experimental conditions do not approach the parameters that define it, the regime in which the attenuation is inversely proportional to the grain size [179].

Elastic scattering by grains is the dominate source of acoustic attenuation in metals, but there are contributions from other causes. They include losses due to inelastic dissipation of energy, absorption, beam spreading and mode conversion [178, 180]. Ultrasonic attenuation in metals by electron relaxation [181] can be included among those causes of energy dissipation not attributed to metallurgical features. Obviously, beam spreading and mode conversion are not fundamental properties of the material, but are caused by geometric and morphological characteristics, and their interaction with the characteristics of the acoustic beam, the transducer and its coupling geometry and their relationship with the specimen.

These non-material contributions to the decay of beam intensity with distance are included in attenuation because they combine with the attenuation caused by material properties to render the reduction observed in the transmitted or reflected signal. Interface characteristics cause the loss in signal amplitude experienced by each successive reverberating echo when it is reflected at the interface between the specimen substrate and the couplant layer connecting the plastic delay line of the transducer assembly to the specimen at interface 0, as illustrated in Fig. 3-1. Although these reductions in signal amplitude are not due to attenuation in the material, but caused by the coefficients of reflection and transmission at that surface, geometric beam spreading and other phenomena, they must be recognized and accounted for in all calculations of reverberating echo amplitudes.

3.2.3 Transducer pulse excitation and response characteristics

Transducer beam anomalies arising from pressure fluctuations in the near field are minimized, and resolution maximized, by the use of short excitation pulses of less than 0.12 μs. A transducer excitation pulse of about 0.12 μs is shown for the pulser used during this investigation in Figs. 5-1 and 5-2 of chapter 5, thus meeting the short-excitation-pulse criterion. The characteristic transducer response and highly damped ring-down are shown in Fig. 3-3(a), a time-response wave form showing transducer response, and Fig. 3-3(b), the resulting envelope of the pulse received by the two 20 MHz transducers.

The envelope shown in Fig. 3-3(b) was derived from the response data provided by the transducer supplier [182] for the two 20 MHz transducers used in this investigation, and is described by the exponential envelope function of time

$$E_1 = c_1(\text{Exp}(-q(t-d_1)^2) + c_2\text{Exp}(-q(t-d_2)^2) + c_3\text{Exp}(-10q(t-d_3)^2)). \qquad (3-21)$$

The values of the constants that determine the particular function that fits the transducer response data are shown in the figure,

where

c_1, c_2 and c_3 are amplitude factors

d_1, d_2, and d_3 are the times at which the maximum of each component of E_1 occurs, and q determines the sharpness of the envelope function.

A 20 MHz signal constrained by that envelope is shown in Fig. 3-3(c). This 20 MHz wave packet characterizing the transducer response, A_t, is described by the product

$$A_t = E_1 \sin(2\pi(t-t_s)/50), \tag{3-22}$$

where

t is the time,

t_s is the time by which the 20MHz wave packet is shifted from the center of the envelope and

50 is the period, in nano-seconds, of a 20 MHz oscillation.

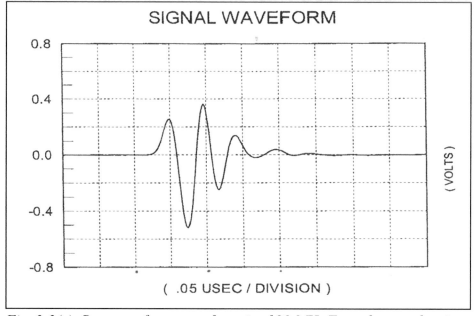

Fig. 3-3(a). Response from one of a pair of 20 MHz Transducers, showing wave form as a function of time (from transducer supplier's data sheet. [182])

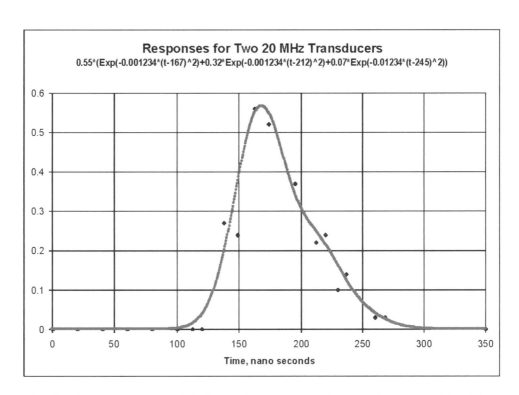

Fig. 3-3(b). Response profile from the two 20 MHz transducers used in this investigation.

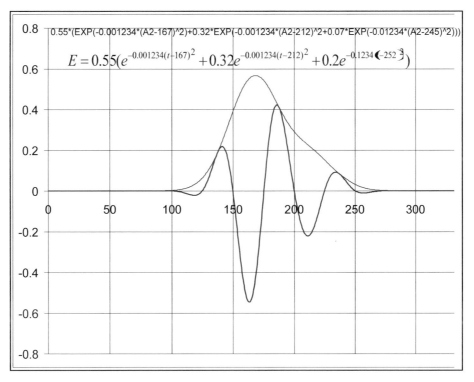

Fig. 3-3(c). A 20 MHz wave form constrained by the transducer response envelope function, E(t).

The following equation, describing successive echo amplitude ratios, shows the contribution from reflection coefficients at both surfaces of a single unbonded substrate sheet, and it shows the attenuation coefficient, a, of the substrate material as a coefficient in the exponent. A schematic illustrating the derivation of the equation for the amplitude of the nth echo, A_n, is shown in Fig. 3-4, where

$$A_n = R_{Ma}{}^n R_{Mc}{}^{(n-1)} A_0 e^{-2anh_M} \qquad (3\text{-}23)$$

and division of equation (3-23), for A_n, by A_{n-1} yields

$$A_n/A_{n-1} = R_{Ma} R_{MC} e^{(-2ah_M)} \qquad (3\text{-}24)$$

where

A_n is amplitude of nth echo exiting the reverberation,

A_{n-1} is amplitude of the previous echo,

R_{Ma} is reflectance at the metal-air interface 1,

R_{MC} is reflectance at metal-couplant interface 0,

a_M is the attenuation coefficient of the metal for the frequency distribution characteristic of the 20 MHz pulse, and

h_M is the thickness of the metal specimen layer interrogated.

Mnemonic alpha characters have been chosen as subscripts over numeric subscripts, in these and subsequent equations, to facilitate reader-friendliness.

Equation (3-23) shows that the reverberation echo amplitude decay is exponential, as expected, and equation (3-24) shows that successive amplitude ratios, A_n/A_{n-1}, are constant for each specimen-couplant combination. These successive amplitudes ratios are valid measurements that are constant for each exponential amplitude decay curve obtained for every combination of reflection coefficients, materials and thicknesses encountered, as long as the attenuation coefficient, a, is constant for the frequency distribution characteristic of the two 20MHz transducers used. An example of this is illustrated by the oscillogram shown in Fig. 3-5(a), resulting from the pulse-echo

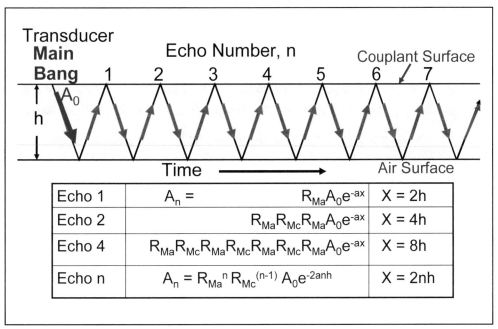

Fig. 3-4. Successive echoes in a single layer

interrogation of a 1-mm thick steel specimen. An exponential decay curve superimposed on peaks of the reverberating echoes in Fig. 3-5(a) is shown in Fig. 3-5(b). The equation shown for the exponential decay curve is of the form

$$A_n = A_{(n-1)} R_{Ma}R_{MC} e^{-an}, \qquad (3-25)$$

where

$A_{(n-1)}$ is the amplitude of the preceding echo,

n represents thickness $2h_M$, as the signal passes through the material twice for each echo, and

a is the attenuation coefficient, a_M, in the metal.

Obviously, equations (3-23) and (3-25) are equivalent, where 2h is represented by n.

The attenuation coefficient in the metal substrate, a_M, can be calculated by (3-24) when the values of the ratio A_n/A_{n-1}, the product $R_{Ma}R_{MC}$, and the thickness, h_M, are known, by

$$a_M = \ln((A_n/A_{n-1})/(R_{Ma}R_{MC}))/-2h_M. \qquad (3-26)$$

A graphic illustration showing how values for a_M can be estimated from A_n/A_{n-1} ratios that were plotted for thin 1020 steel sheet is shown in Fig. 3-6. The scatter is attributed to variation in grain size and orientation concomitant with variations in ingot casting and sheet rolling conditions from one batch to another, and within a given batch. The data plotted at a steel thickness of 0 mm represents the product of reflection coefficients, $R_{Ma}R_{MC}$. The a_M value used for this material is 0.01567/mm, the derivative of the middle curve, and is a more reliable value than would be obtained from any one set of data in equation (3-26), because the curve was produced by a fit to all of the data in the figure.

Acoustic attenuation in the adhesive layer, as in other polymeric materials, cannot be attributed to grains, because these materials are composed of large molecules made from many (poly) molecules (mers) that are chemically linked by carbon-carbon or silicon-silicon bonding during an addition or condensation polymerization reaction. The polymers addressed in this paper will be from the family of those plastic engineering materials that are formed by polymerizing hydrocarbon molecules into long chains that are sometimes more than 500 mers long. These polymers are of interest because they play key roles in developing engineering materials for transportation vehicle construction. Although rubber is a cross-linked thermoset elastomeric polymer, it will not be addressed herein because, despite its important to transportation vehicles, it is not used as an engineering material for vehicle body construction.

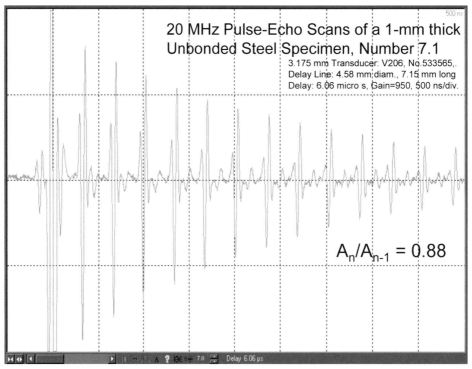

Fig. 3-5(a). Successive echoes from a 20 MHz pulse reverberating in a single steel sheet 1 mm thick

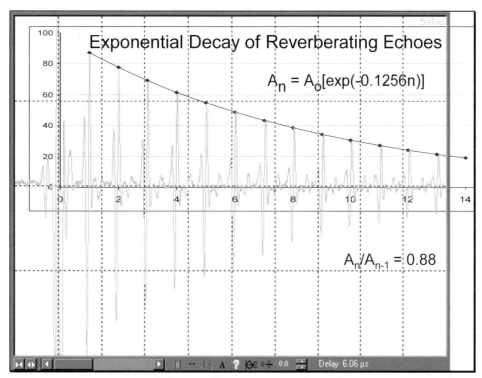

Fig. 3-5(b). Exponential decay of successive echoes from a 20 MHz pulse reverberating in a single steel sheet 1 mm thick.

The acoustic attenuation and velocity in these more than 12,000 polymers and polymer-matrix composites vary, as expected, with their chemical composition, physical characteristics and mechanical properties, and will therefore be addressed as each specific material is encountered in the investigation. Nevertheless, it is noteworthy to observe that the viscoelastic property of these materials contribute to their high acoustic attenuation.

3.3 Reflection of longitudinal ultrasonic waves in layered media

The interrogation of materials by the ultrasonic pulse-echo technique requires the reflection of the pulse from an interface between two transmitting media with different acoustic impedances. Significant differences in acoustic impedances occur at the interface between metal sheets and the adhesive joining them to form a bond joint. The transmission and reflection of the ultrasonic pulse at such interfaces are illustrated for the general case in Fig. 2-7. The specific case for this investigation is illustrated in Fig. 2-8,

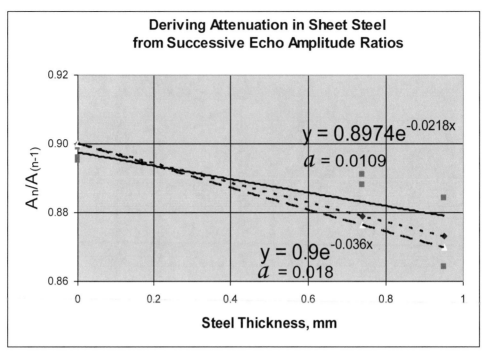

Fig. 3- 6. Curves to Estimate attenuation in 1020 steel sheets less than 1 mm thick

where the pulse is introduced perpendicular to the surface, thus simplifying the expressions for reflection and transmission coefficients. The acoustic impedance mismatch required for a pulse to reflect an echo is shown as the difference between the acoustic impedance values in the numerator of (3-27), the equation used to calculate the reflection coefficient for an interface at a boundary between materials 1 and 2. This coefficient, R, is computed for echo amplitudes [175], for an angle of incident of 0, by

$$R = (Z_2 - Z_1)/(Z_2 + Z_1), \qquad (3\text{-}27)$$

where

Z is the acoustic impedance

subscripts 1 and 2 indicate the sequence order of the material through which the pulse passes on each side of the boundary forming the interface.

Z for each material is the product of the acoustic velocity in the material and the density of the material. Acoustic impedance values are shown for selected materials in table 2-1.

The equation for the reflection coefficient for power, P, for a perpendicular angle of incident is obviously

$$P = ((Z_2 - Z_1)/(Z_2 + Z_1))^2. \qquad (3\text{-}28)$$

However, echo amplitude and phase are of primary interest during this investigation; therefore equation (3-27) will be used as the equation for computing values for R. This will allow R to have positive and negative values, depending on the sequence order of which material the signal traverses on its way to the interface and which material is on the other side of the interface.

The materials involved in this investigation are metals, adhesives, couplant and air. The mnemonic subscripts identifying the acoustic impedances for each of these materials, and concomitant reflection coefficient calculated from them, are shown in equations (3-29) through (3-34) by the mnemonic subscripts M, A, C and a, respectively. The reflection coefficients, R, for each interface are then given by

$$R_{Ma} = (Z_a - Z_M)/(Z_a + Z_M), \text{ at metal-air interface} \qquad (3\text{-}29)$$
$$R_{MA} = (Z_A - Z_M)/(Z_A + Z_M), \text{ at metal-adhesive interface} \qquad (3\text{-}30)$$
$$R_{AM} = (Z_M - Z_A)/(Z_M + Z_A), \text{ at adhesive-metal interface} \qquad (3\text{-}31)$$
$$R_{Aa} = (Z_a - Z_A)/(Z_a + Z_A), \text{ at adhesive-air interface,} \qquad (3\text{-}32)$$
$$R_{CM} = (Z_M - Z_C)/(Z_M + Z_C), \text{ at couplant-metal interface} \qquad (3\text{-}33)$$
$$R_{MC} = (Z_C - Z_M)/(Z_M + Z_C), \text{ at metal-couplant interface.} \qquad (3\text{-}34)$$

where

R_{Ma} is the reflection coefficient at the metal-air interface,

R_{MA} is the reflection coefficient at the metal-adhesive interface,

R_{AM} is the reflection coefficient at the adhesive-metal interface,

R_{Aa} is the reflection coefficient at the adhesive-air interface

R_{CM} is the reflection coefficient at the couplant-metal interface.

R_{MC} is the reflection coefficient at the metal-couplant interface

Acoustic impedance is the product of density and velocity; therefore both the magnitude and sign of R determine the characteristics of the reflected echo returning from each interface. For example, R_{MA} and R_{AM} have the same magnitude, but opposite sign, thus the echo from a pulse that traveled through metal, and was reflected from a metal-adhesive interface, will have an opposite phase characteristic from that of the echo from a pulse that traveled through adhesive and was reflected from an adhesive-metal interface. The physics underlying this phenomenon is explained by an examination of the mechanism of acoustic reflection. Reflections occur when there are differences in the acoustic impedances, Z, of the adjoining materials forming the interface. When the wave approaches the interface through a higher-Z material than is encountered on the other side of that interface, the reflection coefficient, R, is then negative, and "the reflected pressure pulse amplitude has the opposite sign of the incident wave everywhere" [175], so phase inversion occurs because R<0, as can be seen in a variety of reflections observed. Rigorous mathematical treatments supporting the physics of the reflection of acoustic waves are presented by Denisov [174], Schmerr [175] and in more detail by Maev, Titov and Chapman [183].

The general asymmetry of the echoes reverberating in the metal can be attributed to the asymmetry of the impulse response of the transducer, as can be seen in Fig. 3-3(a), and is evident in the transducer envelop function shown in Fig. 3-3(b) and in the simulated wave pattern constrained by that function in Fig. 3-3(c). The negative phase bias exhibited by the first cycle of each echo in the aforementioned Figs. 3-5(a) and (b), is due to the negative reflection coefficient at interface 1, the interface between steel and air. This negative reflection coefficient exists because

$$Z_M >> Z_a . \qquad (3\text{-}35)$$

Similarly, the reflection coefficient, R_{Aa}, at interface 2, the adhesive-air interface, is also negative when the adhesive is not bonded to the metal there, because

$$Z_A > Z_a . \qquad (3\text{-}36)$$

Therefore reverberating echoes from such surfaces will display a prevailing negative asymmetry in their phase when the echo returns from an adhesive-air interface. However, when the echo returns from a bond at interface 2, after passing through the adhesive, R_{AM} is positive and so is the phase of the reflected echo, because $Z_M >> Z_A$.

Now, using the reflection coefficients and the attenuation coefficients, the amplitudes of these reverberating echoes can be expressed as

$$A_n = A_0 R_{Ma}{}^n R_{MC}{}^{(n-1)} e^{-2nha_M} \qquad (3-37)$$

to yield the normalized ratio

$$A_n / A_0 = R_{Ma}{}^n R_{Mc}{}^{(n-1)} e^{-2nha_M} \qquad (3-38)$$

where

A_n is the amplitude of the nth echo

A_o is the initial amplitude introduced into the top layer of the bond joint by the transducer through the delay line and couplant,

R_{Ma} is the reflection coefficient at the metal-air interface, as discussed earlier,

R_{MC} is the reflection coefficient at the metal-couplant interface, as discussed earlier,

n is the echo number

h is the thickness of the material, as indicated by the subscript A or M and

a is the attenuation coefficient of the specimen material indicated by the subscript.

The attenuation coefficient of the plastic delay line, the reflection coefficient at the interface between the delay line and the couplant, the attenuation coefficient of the couplant material, and the reflection coefficient at the couplant-metal interface are all ignored in these calculations because they are not a part of the reflecting and reverberating "system" affected by the bond-joint assembly under inspection, once the acoustic energy has entered the metal specimen material. On the other hand, as discussed earlier, clearly the couplant does participate in damping the amplitude at the metal-couplant interface, interface 1, as the signal returns through the metal and couplant to the transducer. Therefore R_{MC} must be considered as an amplitude-reduction factor at each echo reverberation at the top surface. Quantitative applications of these equations will be illustrated in the next chapter, and the minor impact of the second substrate layer on the adhesive echo analysis, not included in the modeling, will be discussed in chapter 5.

3.4 Propagation of Lamb waves along adhesive bond joints

Acoustic waves can be transmitted in solids by more then seven familiar modes. Much of the previous discussion in this chapter dealt with compression or longitudinal acoustic waves. There are some NDE applications, however, where plate waves offer an advantage. Because their wavelength exceeds the thickness of the solid medium, or plate, along which they are transversely transmitted, plate waves can be propagated over long distances in thin solids. The acoustic field in such a plate medium is comprised of the superposition of incident and reflected bulk waves. These incident and reflected waves must interfere constructively in order for plate waves to propagate. Among these plate waves, Lamb waves are the most commonly used in NDE applications [184, 185]. They are complex-vibration waves in which the particle motion involved in wave propagation exists throughout the entire thickness of a material medium comprising the freely vibrating plate [133]. Lamb wave propagation depends on the density and elastic material properties of the medium, as well as the frequency and material thickness.

Lamb waves are vertically polarized and can be propagated by several modes of particle vibration, but the two most common are symmetrical and asymmetrical. The complex motion of the particles is similar to the elliptical orbits also found for surface waves. The particles propagating symmetrical Lamb waves move in a symmetrical fashion about the median plane of the plate. This is sometimes called the extensional mode because the wave is stretching and compressing the plate in the plane that is transverse to the wave motion direction. Wave motion in the symmetrical mode is most efficiently produced when the exciting force is parallel to the plate. When, however, the Lamb wave is excited by out-of-plane vibrations introduced normal to the plane of the plate, the asymmetrical Lamb wave mode is produced, and often called the flexural mode because a large portion of the motion is in the normal direction to the plate, with a minor component of the motion occurring in a direction parallel to the plate. In this mode, the body of the plate bends as the two surfaces move in the same direction. Further discussion of Lamb waves in the following sub-section will focus on the asymmetrical mode, because that is the mode of propagation used in this investigation.

3.4.1 Velocity of asymmetric Lamb waves propagating along bond joints

The velocity of "flexural" or asymmetric Lamb waves propagating along a plate-type medium is influenced by the stiffness of that plate, derived from what is sometimes referred to as the "flexural modulus" of the material or component. (It should be noted that the "flexural modulus" is not a fundamental material property, but is a measured engineering characteristic caused by shear and tensile modulii.) An expression for the velocity of these asymmetric Lamb waves is developed by Bland [186] and summarized by Schmerr [175] from longitudinal and vertical displacements in x and y, to obtain

$$c = (D_p/\rho h)^{1/4} \omega^{1/2} \qquad (3\text{-}39)$$

where

c   is the wave speed of the fundamental flexural mode,

D_p is the flexural rigidity of the plate,

ρ is the density if the material and

h   is the thickness if the plate.

D_p, the flexural rigidity of the plate, is related to fundamental material properties by

$$D_p = 8\mu(\lambda + \mu)h^3/3(\lambda + 2\mu), \qquad (3\text{-}40)$$

a function of the Lamè constants, λ and μ, comprising components of the material's modulii that constituted equations for longitudinal and shear velocities in (3-11) – (3-18). An examination of the equation for Lamb-wave velocity reveals that the velocity increases as a nonlinear function of modulus and in proportion to the square root of thickness. This relationship between velocity, thickness and stiffness will illuminate the discussion of detecting unbonds with the Lamb-wave NDE technique, because a bonded joint offers significantly more stiffness and thickness than an unbonded joint made from identical material and having the same geometry. Therefore, the presence of an unbonded region reduces the stiffness of the region by effectively reducing the thickness to that of the top layer. The thickness of the plate through which the wave is propagated is reduced because the acoustic energy does not reach the second, or bottom layer substrate, and the velocity of the Lamb wave is reduced, as predicted by the aforementioned equations (3-39) and (3-40). A quantitative treatment is presented in chapter 4 on modeling.

3.4.2 Amplitude of Lamb waves propagating along bond joints

Flexural Lamb waves are excited by the introduction of a vertical vibration at the surface that puts the entire plate into vibration; therefore the amplitude does not diminishes exponentially with depth, but the thickness of the plate makes it stiffer, such that the vibration amplitude does diminish as plates become thicker, because the vibrating excitation force, used by the transducer to introduce the wave into the bonded pair of plates, encounters stiffer resistance when the joint is bonded. This causes the amplitude of the Lamb wave, traveling in a pair of plates that are bonded together, to be much lower than that traveling along an unbonded plate of similar material and geometry. Therefore the amplitude of the detected Lamb wave is also a reliable indicator of the bond state.

4. Analytical Modeling of the Propagation and Reflection of Ultrasonic Waves in Adhesive Bond Joints

This chapter will utilize the mathematical expressions developed in the previous chapter from the physical principles that govern the propagation and reflection of ultrasonic acoustic waves in adhesive bond joints, to develop models that produce outputs which simulate the wave behavior such that they match the experimental results, thereby validating the modeling. The validation of the modeling by experimental results confirms the validity of the mathematical expressions which were used in the modeling and confirms the correctness of theoretical basis from which the expressions were derived. After the validity of the model is confirmed by experimental results, the model can then be used as a predictive tool to identify interferences from unknown anomalies, to correct for interferences from known sources, and to help determine the range of applications over which the ultrasonic techniques are valid. Modeling of both the 20 MHz pulse-echo technique and the 25 kHz Lamb-wave technique will be presented in this chapter.

4.1 Modeling 20 MHz pulse-echoes in sheet-metal adhesive bond joints

Modeling the pulse-echo reverberations in the bond joint was undertaken to demonstrate the validity and applicability of the equations that were derived from an analysis of the physics of the 20 MHz pulse-echo acoustic transmission and reflection process. The modeling will use these analytical equations to simulate the A-scan output from acoustic signals that propagate, reflect and reverberate in the layered bond-joint structure. The model simulates a series of reverberating echoes that result from an acoustic pulse that originated in a 20 MHz piezoelectric transducer, is transmitted through a plastic delay line, then a liquid couplant and into the specimen, where it travels to an interface from which it is reflected and a fraction transmitted, then the reflected echoes reverberate between interfaces, while the amplitude of each successive reflection decays exponentially.

The transducer is equipped with a plastic delay line to provide ample time for transducer damping (about 6 μs) after the excitation pulse is sent, so that the transducer will have completed its "ring down" of post-pulse vibrations, and thereby have enhanced sensitivity to detect the returning echoes. An example of this ring-down is shown by a A-scan in Fig. 4-1, where the first reverberating echo from the metal-air interface of a 1.92-mm thick steel sheet rings down for 9 cycles, or 0.45 μs.

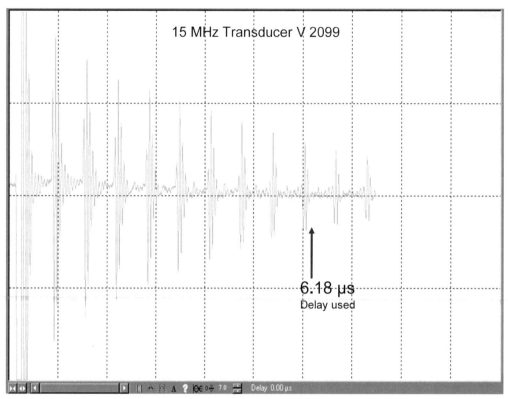

Fig. 4-1. A-scan of 1.92 mm-thick steel sheet with no delay line and no delay showing decay of post-excitation oscillations and residual bias. Gain at 350.

The A-scan also shows the residual bias that is concomitant with no delay. The bias diminishes with time, to nearly zero after 6 µs . Therefore the delay avoids that 0.45 µs ring-down time period, and the 6 µs bias-down time period after transducer excitation. Subsequent echoes reverberate in the top metal layer of the joint structure because a significant fraction of the pulse is reflected at the metal-adhesive interface 1, while a complimentary fraction is transmitted through it. Similar pulse transmissions and echo reflections occur at each interface of the bond joint at which echoes arise, and from which they return to be detected by the sending transducer. Reflections from the bonded adhesive-metal interface 2 are identified by arrows in the A-scan of a fully bonded lap joint shown in Fig. 2-9.

Modeling limits were adopted to exclude those ultrasonic pulse-echo reflections that do not have a direct impact on the analysis and interpretation of signals necessary for adhesive bond joint evaluation. The attenuation coefficient of the plastic delay line, the reflection coefficient at the interface between the delay line and the couplant, the attenuation coefficient of the couplant material, and the reflection coefficient at the couplant-metal interface (exterior to the specimen) are all excluded in these modeling calculations, because they are not a part of the attenuating, reflecting and reverberating "system" comprising the bond-joint assembly under inspection, once the acoustic energy has entered the specimen material at interface 0. The couplant does participate in damping the amplitude of each reflection returning from the metal-couplant interface, as the signal returns through the metal and is reflected at interface 0, while the complimentary fraction is transmitted through the couplant to the transducer. Therefore R_{MC} must be considered at each echo reverberation reflected at the top surface. Slower moving side lobes that may be reflected from the walls of the delay line were minimal in the two transducers used and are excluded from the model. Also excluded are the late and early-arriving echoes that can be attributed to mode conversion and spurious reflections within the transducer-specimen system. These are apparent in the experimental A-scans of unbonded single-sheets as well as bonded specimens. Some occur after the second echo, and grow in amplitude with each reverberation, thus indicating contributions from constructive interference. The most significant of these echoes always manifests a positive phase; therefore they are attributed to reflections from an interface where R > 0. These are excluded from the modeling because they do not impact the echo analysis or interpretation required for adhesive bond joint evaluation.

4.1.1 Echoes reverberating in a single unbonded metal sheet

A model to replicate the echoes reverberating in a single metal sheet was developed to simulate the condition encountered in actual NDE applications when a bond joint is made without the adhesive adhering to the metal surfaces at interface 1. This type of unbonded condition is illistrated at location A of Fig. 2-6, and is equivalent to the pulse-echo examination of a single sheet sketched in Fig. 4-2, with the concomitant reverberating echoes illustrated in Fig. 3-4. To obtain simulated A-scan output from modeling this bond state, and each subsequent bond joint condition, a function that describes the transducer response characteristics is required.

This function was obtained by transforming the sinusoidal wave form representing the transducer response data provided by the transducer supplier [182] and shown in Fig. 3-3(a), to an exponential function of time that defines the transducer response envelop. The resulting compound exponential curve that fits the data shown for the two 20 MHz transducers used, provides the exponential function of time that defines the transducer response envelop shown in Fig. 3-3(b). This transducer response envelop function was optimized by smoothing, shifted in time, then normalized for use as a factor, and is displayed graphically as the compound exponential curve shown in Fig. 4-3.

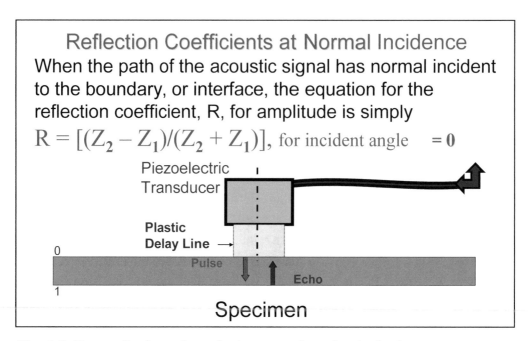

Fig. 4-2. Perpendicular pulse-echo interrogation of a single sheet

This transducer response envelop function, *E*, will be used as the transducer-response envelope defining the time-domain profile of the wave packet reflected from interfaces with the metal substrate. It defines the signal reception profile at each time where a reverberating echo occurs. The function is defined as

$$E = 0.7\text{Exp}(-0.0012345(t - n(2h_M/v_M) - 64)^2) +$$
$$0.35\text{Exp}(-0.0012345(t - n(2h_M/v_M) - 78)^2) \qquad (4\text{-}1)$$

where

t is the time since the main bang introduced the ultrasonic pulse into the specimen

n is the echo number of the pulse reverberating in the metal, and also

n is the number of the echo from the adhesive, arising from the nth echo in the metal

h_M is the metal thickness (mm)

v_M is the velocity of the ultrasonic pulse in metal (0.00589 mm/ns for 1020 steel)

the constant, 0.0012345, is the "q" that determines the sharpness of the envelop function,

the constants, 64 ns and 78 ns, determine the location of each component's maximum

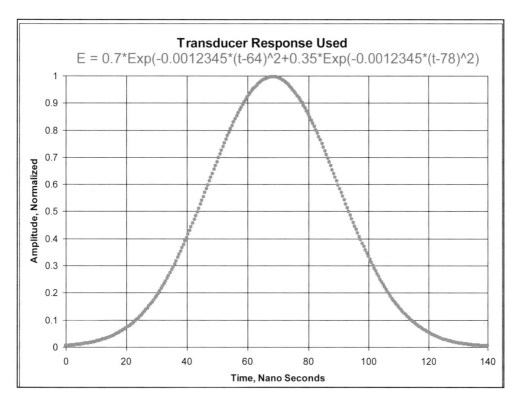

Fig. 4-3 Normalized 20 MHz Transducer response envelop

The transducer-response envelop is only one of the five known factors that determines the amplitude of the 20 MHz signal waveform at each time where a reverberating echo occurs. The other factors include

A_0 , the amplitude of the waveform introduced into the specimen

R_{MA}^n , the reflection coefficient at the metal-adhesive interface, or

R_{Ma}^n , the reflection coefficient at the metal-air interface, used n times,

$R_{MC}^{(n-1)}$, the rerflection coefficient at the metal-couplant interface, used n-1 times,

$Exp(-a_M n2h_M)$, for the attenuation in the metal, or

$Exp(-a_A n2h_A)$, for the attenuation in the adhesive, and

$sin(2\pi(t-n(2h_M/v_M))/50)$, for the 20 MHz frequency of the waveform in metal, or

$sin(2\pi(t-n(2h_A/v_A))/50)$, for the 20 MHz frequency of the waveform in adhesive.

These factors, combined with the envelope function, E, yield the time-dependent amplitude of the 20 MHz signal, A, seen in the simulated oscillographs for each condition modeled. Because this amplitude decays exponentially as a function of n, the echo reverberation number, it will be designated as A_n in the time-dependent equations for echo amplitudes. Moreover, the model must include the reflection coefficients at the metal-air interface, R_{Ma}, when air is present there, as is the case when the metal substrate is unbonded at interface 1. The equation of the waveform for the echoes reverberating in the unbonded metal, for n equal to or greater than 1, is

$$A_n = A_0 R_{Ma}{}^n R_{MC}{}^{(n-1)} Exp(-a_M n 2 h_M) E \sin(2\pi(t - n(2 h_M / v_M))/50), \qquad (4-2)$$

where each constant in the equation is used as defined in the preceeding list of factors.

Attenuation coefficients for the materials involved must be provided in the equations in order for the modeling to produce realistic A-scan simulations. These coefficients were obtained experimentally from data acquired during the laboratory experiments discussed in chapter 5, but are used here in the equations developed for modeling.

Simulated oscillographs resulting from this equation being applied to acoustic echoes reverberating in a material with the mechanical characteristics of 1020 steel sheets are shown for various thicknesses in Figs. 4-4, 4-5, and 4-6, to show the response of the model to changes in thickness over the range of interest. These outputs are validated by oscillograms acquired from A-scans on 1020 steel specimens of similar thicknesses shown in chapter 5. Note that the model and the validating A-scans both show the negative phase assymetry concomitant with phase inversions of reflections from an interface where R<<0, as discussed in chapter 3 on theory. A general expression for the this time-dependent signal, $s(t)$, can be written for any frequency, f_0, and phase shift, $\varphi(t)$, as

$$s(t) = E(t) \sin(2\pi f_0 t + \varphi(t)) \qquad (4-3)$$

Modeling the pulse-echo A-scan should also include a function to yield a valid simulation of the echo from the main bang. This is achieved by assigning a value to A_0, the initial amplitude inside the specimen, so that for n=0, the equation for the amplitude of the main bang, A_{MB}, is

$$A_{MB} = -(A_0/(1 + R_{MC})) R_{MC} (Exp(-a_M n 2 h_M) E \sin(2\pi(t - n(2 h_M / v_M))/50). \qquad (4-4)$$

The amplitude of the main bang, A_{MB}, is calculated from the value assigned to A_0 inside the specimen, after the energy has been transmitted through the couplant-metal interface where the reflection coefficient is R_{CM}, or $- R_{MC}$, hence the negative sign in front of the first term.

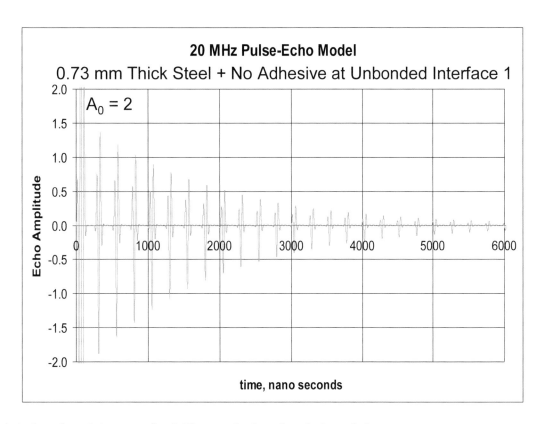

Fig. 4-4. Simulated A-scan of a 0.73 mm-thick unbonded steel sheet.

4.1.2 Echoes reverberating in a single metal sheet coated with a rough adhesive layer

Modeling to simulate the A-scan echoes reverberating in a single metal sheet coated with a rough adhesive layer was done to represent the condition encountered in actual NDE applications when a bond joint is made with adhesive initially adhering to both joined metal surfaces at interfaces 1 and 2, but later pulled apart by an unintended, small separation of the metal sheets before the adhesive is cured, or in some cases afterwards, but likely before cooling. This cohesive separation of the adhesive introduces a rough, non-reflecting unbond within the adhesive layer, with adhesive still adhered to both metal sheets. This type of unbond, resulting from a cohesive failure condition is illustrated at

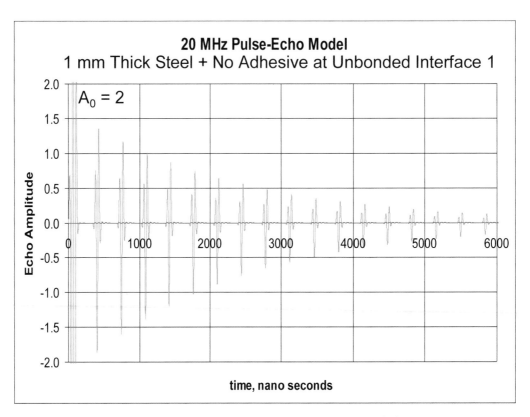

Fig. 4-5. Simulated A-scan of a 1.0 mm-thick unbonded steel sheet.

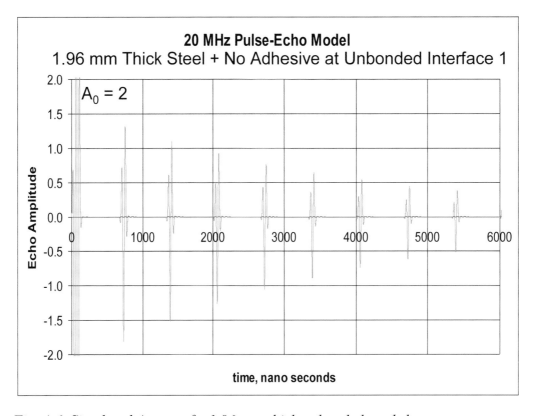

Fig. 4-6. Simulated A-scan of a 1.96 mm-thick unbonded steel sheet.

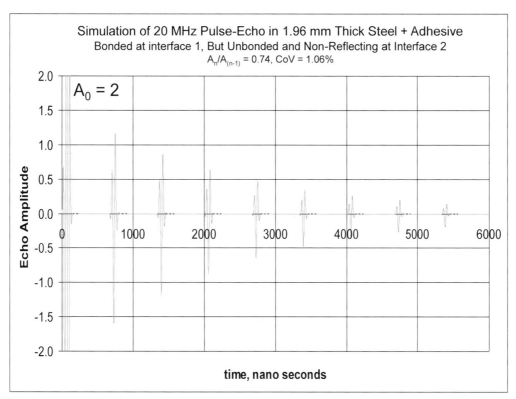

Fig. 4-7. Simulated A-scan of a 1.96 mm-thick steel sheet bonded to an adhesive layer with a rough surface that does not reflect at interface 2.

location B of Fig. 2-6. This situation can also be represented by a metal sheet coated with an adhesive layer that has a rough, non-reflecting outer surface at interface 2.

This unbonded situation results in a reflectance of R_{MA} at the metal-adhesive interface, shown in the schematic of Fig. 3-1 as interface 1, but no reflection from the second interface. The equation corresponding to this situation is

$$A_n = A_0 R_{MA}^{n} R_{MC}^{(n-1)} Exp(-a_M n2h_M)E \sin(2\pi(t-n(2h_M/v_M))/50), \qquad (4-5)$$

where the reflection coefficient, R_{MA}, for the metal-adhesive interface replaces the coefficient for the metal-air interface. Because Z for the adhesive is much less than Z for the metal, the R_{MA} is still negative, and consequently the phase remains inverted as it was for reflections from the metal-air interface. However, the amplitudes of the multiple reflections decline quickly, as can be seen in Fig, 4-7, with succesive echo amplitude ratios of 0.74 for a steel thickness of 1.96 mm.

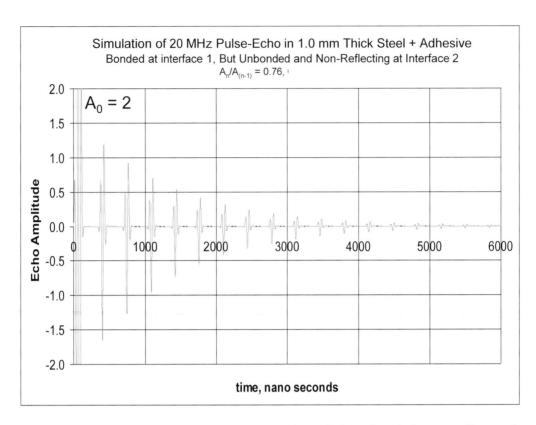

Fig. 4-8. Simulated A-scan of a 1.0 mm-thick steel sheet bonded to an adhesive layer with a rough surface that does not reflect at interface 2.

This rapid decline in successive echo amplitudes is sometimes called "attenuation" for ease of communication, but much of the rapid decline is not due to the material property of attenuation, such as the attenuation in the metal and/or adhesive, but because $R_{MA} < R_{Ma}$, and more of the signal is lost by transmission into the adhesive layer. This significant increase in the "attenuation" of the reverberating echoes provides a robust and reliable method of detecting disbonded regions where adhesive is bonded to the metal at interface 1, but the rough, unbonded opposite surface of the adhesive layer cannot reflect sound at interface 2. A simulation for a thinner steel sheet that is 1 mm thick, coated by an adhesive layer with a rough outer surface, is shown in Fig.4-8, to show how the simulated A-scans respond to change in metal thickness under these conditions.

4.1.3 Echoes reverberating in a single metal sheet bonded to a smooth adhesive layer

Modeling to simulate A-scan echoes reverberating in a single metal sheet, bonded to an adhesive layer with a smooth, reflecting surface at interface 2, was done to correspond to the condition encountered in actual NDE applications when a bond joint is made having adhesion at the first metal-adhesive interface, but no adhesion between the metal and the adhesive layer at the second metal surface, thus reflecting an echo at interface 2. This type of unbonded condition is illustrated at locations C and D of Fig. 2-6. An equation to describe the acoustic pulse transmission and echo reflections concomitant with such a bond state can be developed by adding simulated A-scan echoes from the smooth, reflecting, bonded metal-adhesive interface 1, described by signal echo amplitude equation (4-5), to primary echoes reflected from the adhesive-air interface 2, where there is no bond, then adding those primary echo amplitudes to the secondary and succeeding echoes resulting from the primary reflection reverberating within the metal layer, between the metal-couplant interface at 0 and the metal-adhesive interface at 1. This equation (4-5) must be augmented by a recursion formula that accounts for the added amplitudes from the adhesive echoes returning from the unbonded adhesive-air interface at 2, plus those from previous reflections that are still reverberating in the metal, and adding constructive interference from all past echoes returning through interface 1, from the adhesive-air interface 2.

To derive the necessary recursion formula, let A_{mn} represent the echo amplitude envelop for primary adhesive echoes returning directly from the adhesive-air interface 2, through the metal sheet, on to which their secondary and successive reverberations will be added to subsequent similar primary echoes, so that

$$A_{mn} = A_{11}, A_{12}, A_{13}, A_{14}, A_{15} \tag{4-6}$$

where $m = 1$ and $n = 1, 2, 3, 4, 5$, etc, to $n = \infty$

and

$$A_{adh} = n > 0, (A_0 \mathrm{Exp}(a_M 2nh_M) ER_{Aa} R_{MA}^{(n-1)} R_{MC}^{(n-1)} (1 - R_{MA})(1 - R_{AM}) \mathrm{Exp}(-a_A 2h_A)). \tag{4-7}$$

And subsequent adhesive echo amplitudes are added to the primary adhesive echoes, using a materials properties function, G, for simplification, where

$$G = R_{MA} R_{MC} \mathrm{Exp}(-a_M 2h_M) \tag{4-8}$$

such that

for n=2:

$$A_{21} = A_{11} G \tag{4-9}$$

for n=3:

$$A_{22} = A_{12} G, \; A_{31} = A_{21} G \tag{4-10}$$

for n=4:

$$A_{23} = A_{13} G, \; A_{32} = A_{22} G, \; A_{41} = A_{31} G \tag{4-11}$$

for n=5:

$$A_{24} = A_{14} G, \; A_{33} = A_{23} G, \; A_{42} = A_{32} G, \; A_{51} = A_{41} G, \tag{4-12}$$

and so on, so that for the total adhesive echo amplitude, A_{total}, at any n^{th} arrival is

$$A_{total} = \sum_{m=1, n=1}^{m=n, n=\infty} A_{m,n} G \quad \text{over } m = 1 \text{ to } \infty \text{ and } n = \infty \text{ to } 1. \tag{4-13}$$

So that equations (4-5) and (4-13) can be combined, according to the superposition principle, to yield the output oscillographs from "half-bonded" specimens that are shown in the simulated A-scans pictured in Figs. 4-9, 4-10 and 4-11 for materials with characteristics corresponding to metal sheets of various thicknesses. These figures show simulated A-scans from two different metal substrates, with different thicknesses, bonded to several thicknesses of the adhesive used in this investigation. These outputs from the model are validated by oscillograms acquired from A-scans on similarly bonded metal specimens of similar thicknesses shown in chapter 5. Note that the model and the validating A-scans both show the negative phase assymetry concomitant with phase inversions of reflections from an interface where R<0.

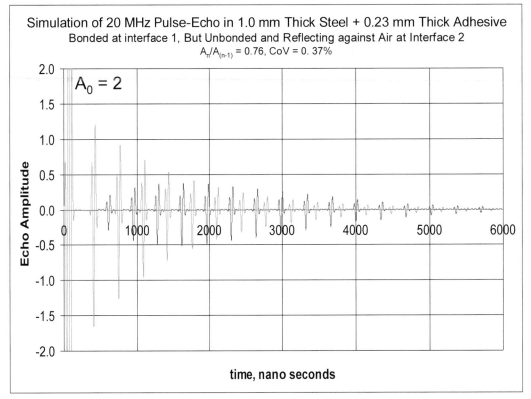

Fig. 4-9. Simulated A-scan of a 1.0 mm-thick steel sheet bonded to a 0.23 mm-thick layer of adhesive with a smooth, reflecting, unbonded surface at interface 2.

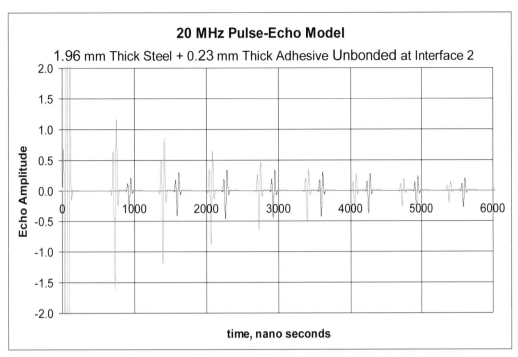

Fig. 4-10 . Simulated A-scan of a 1.96 mm-thick steel sheet bonded to a 0.23 mm- thick adhesive layer with a smooth, reflecting, unbonded surface at interface 2.

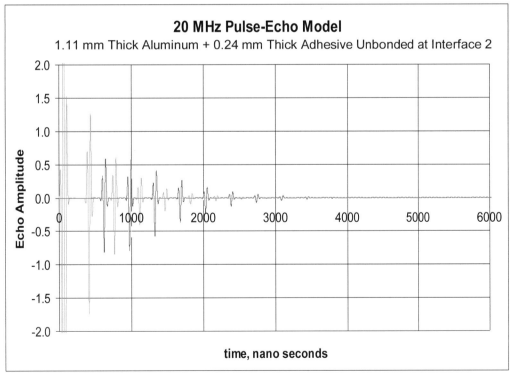

Fig. 4-11 . Simulated A-scan of a 1.11 mm-thick aluminum sheet, bonded to a 0.24 mm-thick adhesive layer with a smooth, reflecting unbonded surface at interface 2. Echoes 2, 4, 6, 8, 10 and 12 are from the adhesive-air reflections at interface 2.

93

The next step in the development of equations that describe the transmission and reflection of acoustic signals in adhesively bonded joints, and modeling their output to confirm the validation of the equations, is to analyze the acoustic transmission and reflection in a bonded metal sandwich composed of two metal sheets bonded to an adhesive at both interface 1 and at interface 2, with reflecting surfaces at each interface.

The negative phase preference observed in Fig. 4-12 is a reliably repeatable phenomenon and was observed in each A-scan of an unbonded joint during this investigation. The contrast between the adhesive echoes shown in Fig. 4-12 and those shown in Fig. 4-13 is noteworthy. The contrast between the negative phase preference of the echoes returning from the adhesive-air interface 2, when no bond exists there, and the positive phase of those returning from the adhesive-metal interface 2, when a bond exists there, is vivid. It is a reliably repeatable phenomenon and was observed in each A-scan of an unbonded or bonded joint during this investigation. It is caused by the change in the sign of the reflection coefficient, as demonstrated clearly by the model simulations run at various thicknesses of both the substrate metals and the adhesive.

An expanded time scale and an increase in the thickness of the metal specimen, by nearly a factor of two, was used to produce both real and simulated A-scan oscillograms to provides a clearer analysis of the aforementioned negative phase preference and the echo from the "main bang" that occurs when the ultrasonic pulse is introduced into the sheet. The expanded time scale also shows more details about the echoes reverberating in the metal specimen. Since the steel used for this specimen was virtually the same as that used to produce many of the other A-scans, the attenuation coefficient, a_M, is virtually the same. The thickness, h_M, of 1.96 mm yields the predicted exponential decay, expressed as the ratio of $A_n/A_{(n-1)}$, for a steel sheet that is 1.96 mm thick, with couplant at he transducer surface, and corresponds to an attenuation coefficient of 0.0175 mm^{-1} .

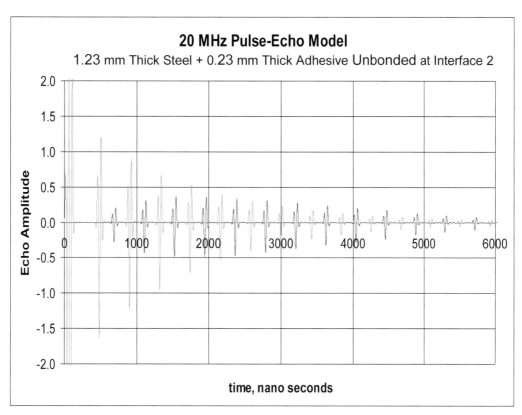

Fig. 4-12. Simulated A-scan of a 1.23 mm-thick steel sheet bonded, to a 0.23 mm- thick adhesive layer with a smooth, reflecting unbonded surface at interface 2.

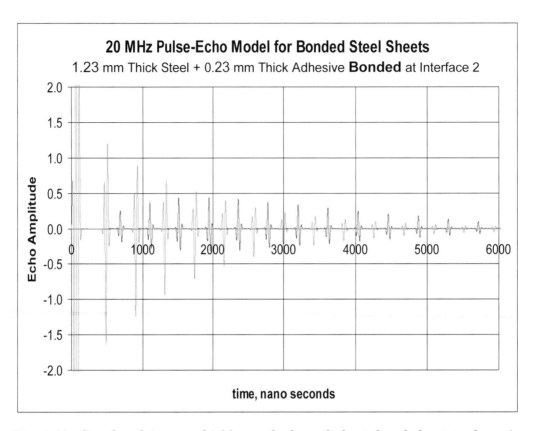

Fig. 4-13 . Simulated A-scan of 1.23 mm-thick steel sheet, bonded at interfaces 1 and 2 to a 0.23 mm-thick adhesive layer.

Note that the magnitude as well as the sign of the reflection coefficient, R, determine the amplitude and phase preference, respectively, of the reflected echo. So when R_{AM} interface 2 is positive, the reflected echo shows a positive phase preference.

To confirm the validity of the approach used to combine primary, secondary and subsequent adhesive echo amplitudes from reflections at the adhesive-air interface 2, when no bond exists there, or the adhesive-metal interface 2, when bond does exist there, ratios of the adhesive echo amplitudes to the steel echo amplitudes were measured for the simulated A-scans and compared th those acquired from the experimental A-scans in chapter 5. These ratios are plotted and shown in Figs. 4-14 and 4-15 for comparison.

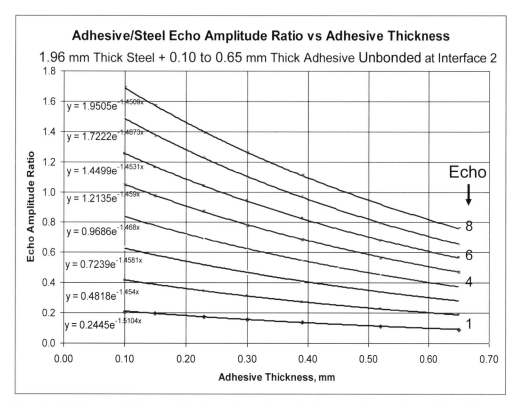

Fig. 4-14. Ratio of the amplitude of simulated echo number n, from the adhesive-air interface at interface 2, to the amplitude of the corresponding simulated echo number n from the steel-adhesive interface at interface 1. Comparing this family of curves with a family of corresponding curves generated from data validates the model.

The agreement between simulated and experimental amplitude ratio data of corresponding echoes from the adhesive layer at interface 2, to the amplitude of the corresponding simulated echoes from the steel sheet at interface 1, is a valid indication of the correct combination of adhesive amplitude ratios.

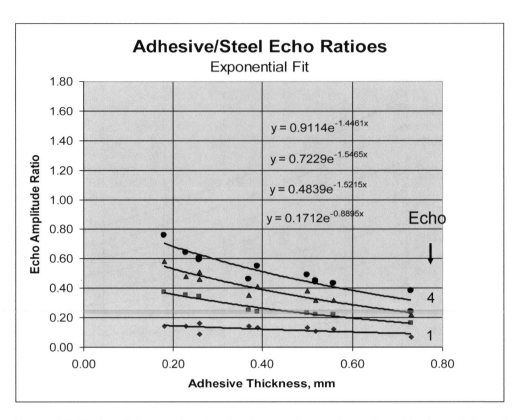

Fig. 4-15. Ratio of the amplitude of echo number n, from the adhesive-air interface at interface 2, to the amplitude of the corresponding echo number n, from the steel-adhesive interface at interface 1. Comparing this family of curves from data with a family of corresponding curves generated by the modeling validates the model. 1.

4.2 Modeling 20 MHz pulse-echoes in plastic and polymer composite joints

A simulated A-scan output from the model is shown in Fig. 4-16 for a polystyrene material with an acoustic velocity, $v_L = 2350$ m/s and attenuation, $a_p = 0.24$, about ⅓ of the attenuation measured in the adhesive. This figure shows an artificially-enhanced amplitude for the returning echoes because the reflection coefficients at interfaces 0, 1 and 2 have been incorrectly set to unity in order to make the echoes appear. This was done so that the reader has a reference that identifies the echo locations in the time-domain. To enhance echo amplitudes, A_0 was increased three-fold, from 2, used in previous simulated A-scan displays, to 6, in order to compensate for the loss of amplitude due to higher attenuation in the polymeric material. Figure 4-17 shows a simulated A-scan with a further enhanced initial amplitude, A_0, set to 20, and with the correct reflection coefficients calculated from the acoustic impedances of the materials joined at each interface. These refection coefficients for each specific interface were calculated by equation (3-26), the same equation shown in Fig. 4-2, and are listed here.

The reflection coefficient at the polycarbonate-couplant interface 0, $R_{PC} = -0.0112$

The reflection coefficient at the couplant-polycarbonate interface 0, $R_{CP} = 0.0112$

The reflection coefficient at the polycarbonate-air interface 1, $R_{Pa} = -0.9997$

The reflection coefficient at the polycarbonate-adhesive interface 1, $R_{PA} = 0.0526$

The reflection coefficient at the adhesive-polycarbonate interface 0, $R_{AP} = -0.0526$

The reflection coefficient at the adhesive-air interface 2, remains $R_{Aa} = -0.9996$.

Using these data in the model, and attenuation values approximating that obtained experimentally from data presented in and discussed in chapter 5 for the adhesive, the simulated A-scan in Fig. 4-17 shows only one detectable first echo reflected from the plastic-adhesive interface 1. This is because the reflection coefficient at that interface indicates that a maximum of less than 5.3 % of the signal can be reflected from there. It is also important to note that the small echo that does appear at 2618 nano-s shows a positive phase preference, because the reflection coefficient at that interface, R_{PA} , is positive. As has been discussed in chapter 3 as well as modeled this chapter, and will be shown by the experimental data presented in chapter 5, the phase preference of the reflected echo is always governed by the sign of the reflection coefficient.

The 10-fold increase in gain, used here in the model to amplify the small echo so that it can be seen at 2618 nano-s, would also increase the noise in real A-scans. This unavoidable noise is due to several sources, including mode conversions, side-lobe signal delays, transducer electronics, and material anomalies. Polymer composites would manifest even greater attenuation and noise than metals or homogeneous plastics, necause in heterogeneous composites, much of the noise is caused by the reflections, scatter and velocity dispersion of the acoustic energy as it is transmitted across impedance-mismatched interfaces between the polymeric resin and the mineral or metal reinforcement strands and/or particles. Obviously, increased noise would adversely impact signal-to-noise ratio, thus marginalizing the ability to detect the weak echoes from the adhesive layer, and would certainly minimize the opportunity to analyze the phase preference used to determine which surface of the adhesive, if any, is unbonded.

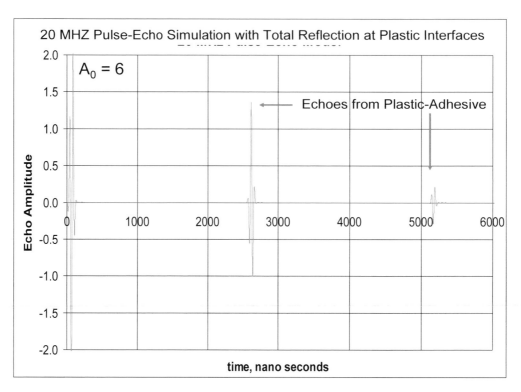

Fig. 4-16 . Simulation of reverberating echoes in a 3 mm thick polystyrene sheet, with A_0 tripled to 6 and reflection coefficients set at 1 in order to show echo times.

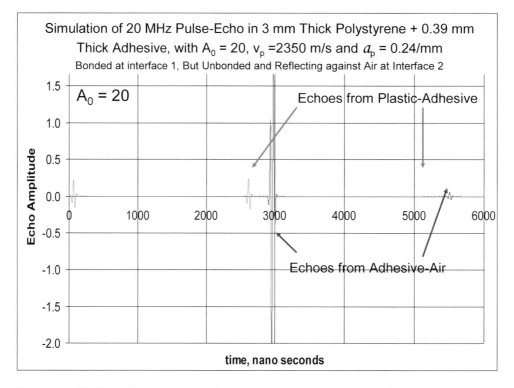

Figure 4-17. Simulation of reverberating echoes in a 3 mm thick polystyrene sheet bonded to 0.39 mm thick adhesive layer, with A_0 increased to 20.

Further insights can be gained from an examination of the amplitudes and phases of echoes in Fig. 4-17. The very low amplitude reflected by the main bang at the left side of the simulated A-scan corresponds to the very low reflection coefficient, R_{CP}, at interface 0, where no more than 1.1 % of the signal is reflected. The phase preference is positive because $R_{CP} > 0$. Conversely, the echoes at the adhesive-air interface 2 show negative phase.

Additional experimental observations that validate the modeling simulation of Figs. 4-16 and 4-17 are shown in Fig. 4-18. Again, the small reflected negative echo shown at the left of the main bang indicates a small $R_{PC} < 0$. The relatively low amplitude, positive phase reflection representing the main bang indicates a small $R_{CP} > 0$. The major echo in Fig. 4-18 is the first echo from the plastic-air interface 1, where the R_{Pa} approaches unity. Subsequent echoes from reverberations in the 0.5 mm plastic plate are quickly diminished by a low value of R_{PC} at interface 0, where reverberating echoes are reflected, and by the high attenuation in the plastic material. The successive amplitude ratio at which these echoes diminish can be calculated by equation (3-22), using applicable constants.

$$A_n/A_{n-1} = R_{Pa}R_{PC}\exp(-2ah_p) \tag{3-22}$$

In this case, $A_n/A_{n-1} = 0.00881$, which means that the amplitude of each echo is only 0.88 % of the amplitude of the preceding echo. Hence the third echo from the plastic-air interface 1 is thoroughly buried in the noise along the axis.

The minimal simulated echoes reflected from the substrate-adhesive interfaces in Fig. 4-17, as well as those with rapidly diminishing amplitude confirmed in the experimental A-scan of Fig. 4-18, do not permit the analysis of a series of several echoes to determine the bond state of the adhesive joint, nor the surface of the adhesive layer where that bond state exists. Moreover, these A-scan illustrations explain why there is a need for the Lamb-wave approach in those cases where high attenuation and small reflection coefficients reduce the substrate-adhesive echoes to nearly noise level, so that the signal to noise ratio falls below the statistical detection limit.

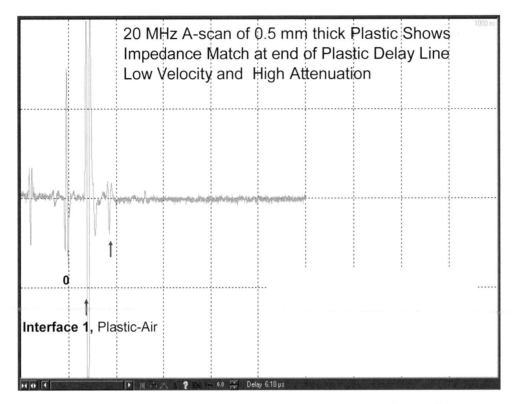

Figure 4-18. A-scan of 0.5 mm thick polycarbonate sheet with no adhesive

4.3 Modeling 25 kHz Lamb waves in plastic and polymer composite joints

Modeling the 25 kHz Lamb wave signal that is transmitted along an adhesively bonded joint and received by a piezoelectric transducer with a resonant frequency of 25 kHz, requires that a profile of the transducer signal-response characteristics be established. This characterization of the 25 kHz transducer was accomplished by experimental observations made while the transducer was acquiring signals on similar material and joint configurations to that to which it would be applied during use. These observations, and the data acquired there from, led to the analytical expression of the transducer response profile, or envelop, expressed in equation (4-14) and shown graphically in Fig. 4-19.

$$E_L = 0.65(Exp(-0.000033(t-350)^2)+0.55Exp(-0.000035(t-360)^2))sin(2\pi(t-22.5)/40) \quad (4-14)$$

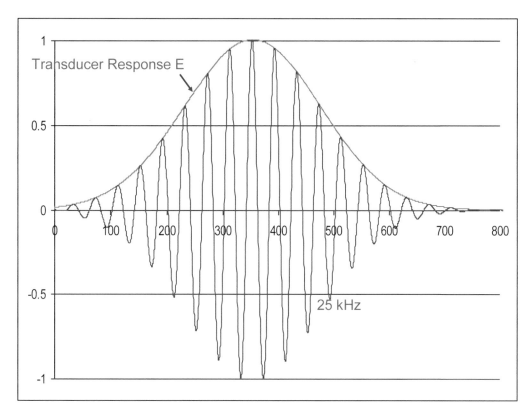

Fig. 4-19. The 25 kHz Lamb-wave transducer response envelope, with the 25 kHz received signal constrained by it.

where

E_L = the 25 kHz Lamb-wave transducer response envelop function of time, and

t = time, in μs

The modeling of Lamb wave propagation along the bond joint, as well as their reception and transduction, uses this exponential envelop function, E_L, which analytically defines the 25 kHz transducer response to received signals propagated during the modeling of bonded and unbonded joints. The Lamb-wave velocity, c, of the fundamental flexural mode can be calculated from the equation introduced in chapter 3 on theory.

$$c = (D_p/\rho h)^{1/4} \omega^{1/2} \tag{3-39}$$

and

$$D_p = 8\mu(\lambda + \mu)h^3/3(\lambda + 2\mu). \tag{3-40}$$

When using the Lamb wave approach, the detection of unbonded regions is based on changes in the 25 kHz wave propagation that occur while interrogating a bond joint of constant material and geometry. Therefore all of the variables in the two aforementioned equations remain constant, except h, the effective thickness of the joint. The presence of an unbonded region reduces the stiffness of the region by effectively reducing the thickness to that of the top layer. The thickness of the plate through which the wave is propagated is reduced because the acoustic energy does not reach the second, or bottom substrate, but is confined to the top substrate layer and to any adhesive that may be bonded to it. This means that (3-39) can be simplified, using (3-40) to define D_p/h^3 as

$$D_p/h^3 = 8\mu(\lambda + \mu)/3(\lambda + 2\mu), \text{ then the velocity,} \tag{4-15}$$

$$c = Kh^{\frac{1}{2}} \tag{4-16}$$

where

$$K \text{ is } (\omega^2 D_p/h^3 \rho)^{1/4}$$

So that solving the above equation for the asymmetric Lamb-wave velocity, c, in a polymeric material with a total bond-joint thickness of 6 mm, and evaluating K where the velocity of the Lamb wave along the bonded joint region is

$c_B = 833$ m/s,

would exhibit a velocity, c_U, along an unbonded region of an identical material with effectively half the thickness of the bonded region, or 3 mm, yields a velocity of

$c_U = 589$ m/s.

This lower velocity causes a delay of 9 µs in the arrival time at the receiving transducer located 18 mm away. The peak shown at 125 µs in Fig. 4-20 would shift to 134 µs when the transducer is moved to an unbonded region of the joint. This calculated shift is confirmed by the actual bond NDE data presented as oscilloscope images in chapter 7, and shown in Figs. 7-5(a) and 7-5(b), where the arrival times of the two wave trains from bonded and unbonded regions are compared.

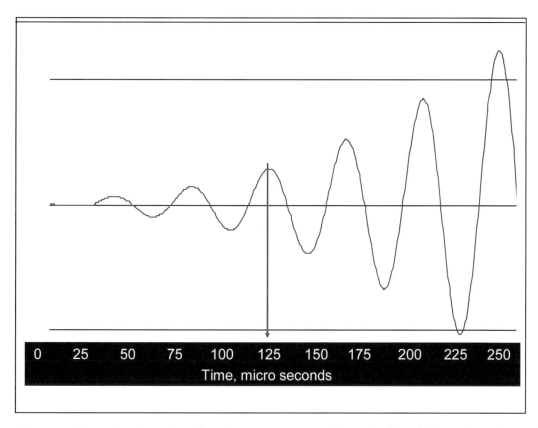

Fig. 4-20. Simulated 25 kHz Lamb wave response from the bond joint of a polymer in which the Lamb-wave velocity is 833 m/s.

The change in Lamb-wave amplitude with plate thickness can be derived by differentiating equation (3-40) while holding the Lamé constants constant, because only the thickness is changing, so that

$$\partial D_p/\partial h = 3D_p/h. \tag{4-17}$$

Furthermore,

$$D_{p(at\ half\ thickness)}/D_p = (8\mu(\lambda + \mu)/3(\lambda + 2\mu))(h/2)^3/(8\mu(\lambda + \mu)/3(\lambda + 2\mu))h^3 = \tfrac{1}{8}. \tag{4-18}$$

Therefore the amplitude of the oscillations will be greater over an unbonded region because the plate, or in this case the unbonded bond joint, is virtually half the thickness and has only $\tfrac{1}{8}$ the flexural rigidity of the bonded region. Of course, the unbonded region may be small and thus restrained from full-flexural excursions by adjacent bonding. Moreover, the driving force and limited translation amplitude available from the transducer at 25 kHz is not expected to move the unbonded near-layer substrate by eight times more than the bonded sandwich, but it is clear that the amplitude of oscillation will be greater. The approximately three-fold increase in the amplitude of the signal shown in Fig. 7-5(b) from an unbonded region, over the amplitude from a bonded region, shown in Fig. 7-5(a), offers experimental evidence that the data verifies the trend predicted by equation (4-18), and shown in the simulated signal displayed in Fig. 4-21.

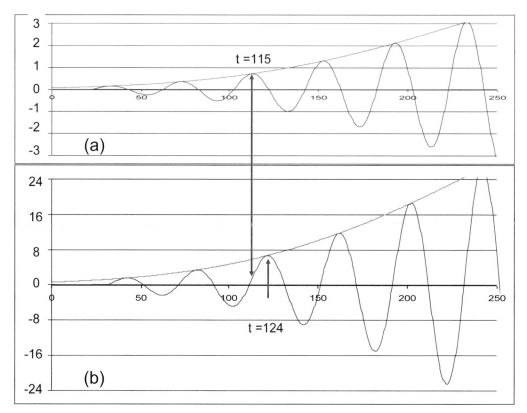

Fig. 4-21. Decrease in Lamb wave velocity along joint, from bonded region, modeled in (a), to unbonded region, modeled in (b). Corresponding increase in amplitude, from bonded region (a) to unbonded region (b).

4.4 Validation of Models by Experimental Results

4.4.1 Validation of 20 MHz pulse-echo model

The 20 MHz pulse-echo model will be validated by the experimental results presented in chapter 5. The A-scan data showing exponential decay due to attenuation in the material and reflection coefficients at the interfaces match the output displayed by the simulated A-scans from the modeling. Moreover, the ratios of echo amplitudes returning from the adhesive to echo amplitudes from the metal-adhesive interface also validate the modeling.

4.4.2 Validation of 25 kHz Lamb-wave model

The 25 kHz asymmetric Lamb-wave model is validated by the experimental results presented in chapter 7.

5. Development of a 20-MHz Pulse-Echo NDE Technique for Bonds

The development of a 20-MHz acoustic pulse-echo method for nondestructive evaluation (NDE) of adhesive bonds was undertaken to provide assurance of bond integrity along bond joints in vehicle body assemblies. As the demand for improvements in fuel-efficiency, performance, corrosion resistance, body stiffness and style increases, the use of adhesives continues to grow to help meet those demands by providing wider design, materials and process options. A history of inconsistent adhesive bond performance in mass-production transportation vehicle applications, however, indicates the need for a robust method of insuring adhesive bond integrity in manufacturing. Therefore this 20 MHz pulse-echo method has been developed to meet this need.

This new bond NDE method provides improvements over previous methods implemented in production, and extends the range of effectiveness for evaluating the bond state, to a resolution of 4 mm, on both near and remote interfaces between the substrate and the adhesive. This degree of resolution for both bond state and interface location has not previously been available for adhesive bonds as thin as 0.12 mm. The method detects the bonded or unbonded state of the first adhesive interface encountered, and if it is bonded, then identifies the bond state of the second by a phase-sensitive technique.

5.1. Principles of operation

The NDE is accomplished by the acquisition and analysis of acoustic echoes that return from the ultrasonic pulses sent into the specimen for the interrogation of bond joints that have interfaces between layers of material with large acoustical impedance mismatch, such as the two metals and adhesive materials shown in table 2.1. These reflected echoes result from an acoustic pulse that originates in an electrically excited 20 MHz piezoelectric transducer, is transmitted through a plastic delay line, then a liquid couplant and into the specimen, where it travels to an interface from which it is reflected, as a fraction is transmitted. The reflected echoes reverberate between interfaces in the multilayered joint structures, while the amplitude of each successive reflection decays exponentially. The echoes are captured for an analysis that provides a clear indication of the bond state, and can ultimately yield a simplified display of the interpretation of the inspection results.

The indications and the resulting interpretations are presented to show how the presence of bonds at the first interface, between the first metal layer and the adhesive, are recognized by the increased attenuation rate of echoes reverberating in the first metal sheet, and by echoes from the second adhesive interface, when that adhesive surface is smooth and reflecting. When the adhesive surface is smooth at interface 2 and reflects an echo, but is not bonded to the metal there, the reflected echo is returned from that interface with a negative phase, because

$$R_{Aa} < 0, \qquad (5\text{-}1)$$

where

R_{Aa} is the reflection coefficient at the adhesive-air interface.

Bond state at the second interface is evaluated by this phase-sensitive analysis of the echoes reflected from that adhesive-metal interface as a consequence of the reflection coefficient,

$$R_{AM} > 0, \qquad (5.2)$$

where

R_{AM} is the reflection coefficient at the adhesive-metal interface.

Therefore the phase of the reflected echo is reversed from negative, when $R_{Aa} < 0$ at an unbonded interface, to positive, when $R_{AM} > 0$ at a bonded interface. This phase-sensitive analysis of the echoes reflected from both interfaces 1 and 2 with the adhesive layer was explained in chapter 3 on theory and applied in chapter 4 on modeling.

5.2 Apparatus and Experimental Methods

The equipment used in these experiments consist of an often-mentioned highly damped 20 MHz transducer, a pulser-receiver and a desk-top personal computer (PC) equipped with software to enable the display of the digitized ultrasonic echo signal on the PC screen as if it were a digital oscilloscope. The PC and software provided the capability to display ultrasonic echo signals as A-scans and B-scans.

5.2.1 Pulser-receiver background contributions

In order to distinguish the signals contributed by the specimens and their anomalies from the background contributed by other sources, several artificial "A-scans" were acquired on background signals and noise to identify background contributions from several sources that are components in the pulse excitation and data acquisition system. The base-line contributions from the electronics, including the pulser-receiver and coaxial cable circuitry, are shown as a quasi A-scan presentation in Fig. 5-1, for gain set at 1000 and without the transducer attached. There is no delay and the time scale is expanded by a factor ten more than that used throughout this investigation, to 0.1 µs/division, in order to resolve time-dependant features. Note the electronic noise concomitant with the delivery of the main bang pulse.

Subsequently, the 20 MHz transducer was attached to the coaxial cable connecting it to the pulser-receiver circuitry, and the resulting background is shown in Fig. 5-2. The high-amplitude excursion concomitant with the delivery of the main bang pulse is virtually unchanged by the presence of the transducer during the initial 0.35 μs, but the subsequent ring-down oscillations, showing a period of 50 nano s, or a frequency of 20 MHz, are clearly characteristic of the attached transducer, and they appear to damp out after about 0.25 μs. This is about the same ring-down time as that indicated by the transducer supplier [182] and shown in Fig. 3-3(a), but longer than the 0.14 μs used in the modeling of chapter 4 and shown in Fig. 4-2. The reason for the difference can be attributed to the degradation of the signal that appears near the end of the ring down in Fig. 5-2. Note that the excitation bias is higher with the transducer present than that shown without the transducer connected in Fig. 5-1.

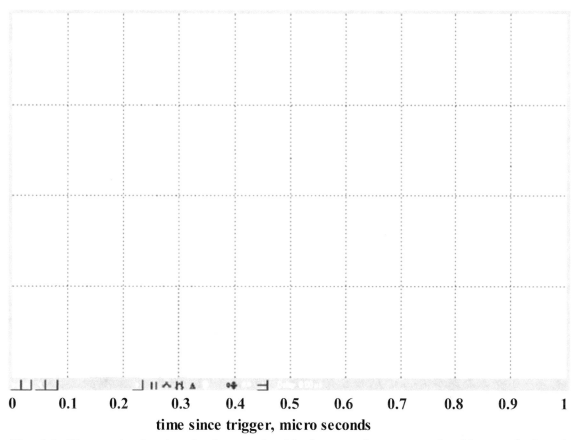

0   0.1   0.2   0.3   0.4   0.5   0.6   0.7   0.8   0.9   1
time since trigger, micro seconds

Fig. 5.1. Electronic circuitry background, with the transducer coaxial cable attached to the pulser-receiver, but with no transducer. Delay: 0, Gain:1000, Time scale: 0.1 μs/division

The time scale is then adjusted to 1 μs/division, that used during actual ultrasonic data acquisition, the gain remained set to the operating level of 1000, and the 20 MHz piezoelectric transducer remained attached to the system to acquired the background that appears in Fig. 5-3. Note the excitation bias that decays in about 6 μs after the main bang. The "main-bang" echo from the interface between the 0.5 mm-thick couplant column and the work piece, or specimen, appears at 0 μs. Subsequent echoes reverberate in the top metal layer of the joint structure, because a significant fraction of the pulse is reflected at the metal-adhesive interface 1, while a complimentary fraction is transmitted through it. Similar pulse transmissions and echo reflections occur at each interface of the bond joint at which echoes arise, and from which they return to be detected by the sending transducer. Reflections from the bonded adhesive-metal interface 2 are identified by arrows in the A-scan of Fig. 2-9.

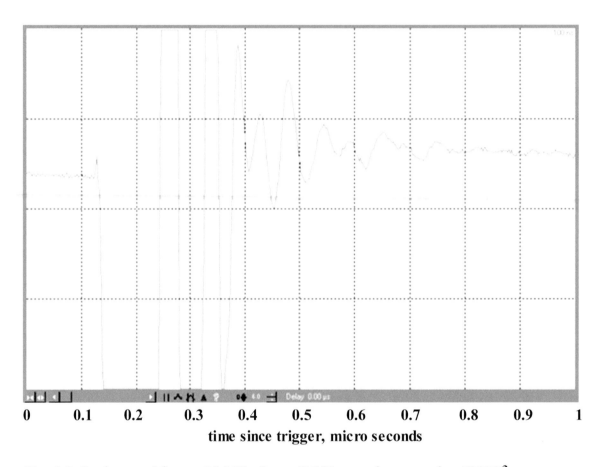

time since trigger, micro seconds

Fig. 5.2. Background from a 20 MHz, 3mm, V 208 transducer number 533583
Delay: 0, Gain:1000, Time scale: 0.1 µs/division

5.2.2 Transducers selection and characteristics

The use of piezoelectric transducers to excite and detect high-frequency ultrasonic waves for the characterization of solid materials is well-known technology and a plethora of NDE literature has reported an abundance of widely varying applications in every industry where materials and processes technologies are essential. Such transducers are key components of one of the most attractive approaches to inspection technologies, because they excite and detect acoustic waves that propagate through the materials in such a way as to provide effective primary indicators of the mechanical and physical properties of the specimen, as well as anomalies that may be present therein. As expected, these advantages do not come without limitations. Some limitations arise from the physics of the material, design, excitation, and operation of the transducer, as well as the coupling of the transducer to the specimen under interrogation. Because of these transducer limitations, and their impact on the excitation and interpretation of the

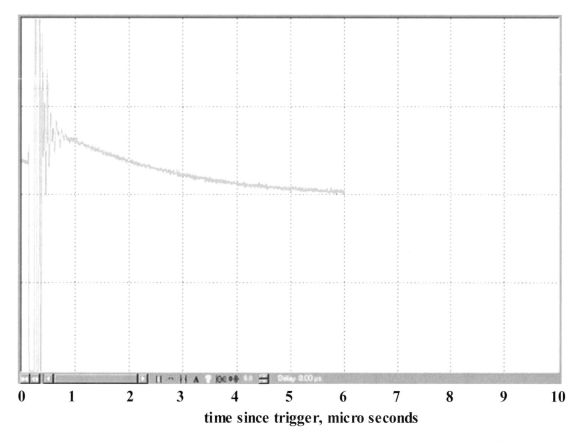

time since trigger, micro seconds

Fig. 5-3. Background from a 20 MHz, 3mm, V 208 transducer number 533583
Delay: 0, Gain:1000, Time scale: 1 μs/division

acoustic signals from specimens under interrogation, the characterization of the transducer is an essential prerequisite to its effective use in quantitative ultrasonic inspection.

The 9-mm outside-diameter transducer probe used in this investigation allows inspection where vehicle geometric constraints prohibit large inspection probes, such as that used in chapter 7 for 25 kHz Lamb-wave propagation. A schematic of the transducer, with plastic delay line and couplant are shown on a bond joint in Fig. 2-1. The highly damped 20 MHz transducer response waveform is shown in Figs. 3-3(a) and 3-3(b). The excitation potential and pulse repetition, supplied by the pulser, are determined by specifically designed circuitry to optimize the performance and sensitivity of the transducer.

The importance of a highly damped transducer is illustrated by contrasting the residual oscillations in an A-scan acquired with a less-damped 15 MHz transducer, shown in Figs. 5-4(a) and 5-4(b), to an A-scan acquired with the highly damped 20 MHz

111

transducer used during this investigation, shown in Fig. 5-5. The residual oscillations are clearly visible following the main bang reflection, and can be seen with diminishing amplitude following each succeeding echo reverberation. These residual oscillations can sometimes interfere with a visual diagnosis of the A-scan for subtle anomalies.

The 20 MHz frequency was selected in order to provide the high resolution desired, because the high frequency produces a 0.115 mm-wavelength in the adhesive, as shown in table 2.1, that is short enough to effectively evaluate adhesive layers as thin as 0.12 mm.

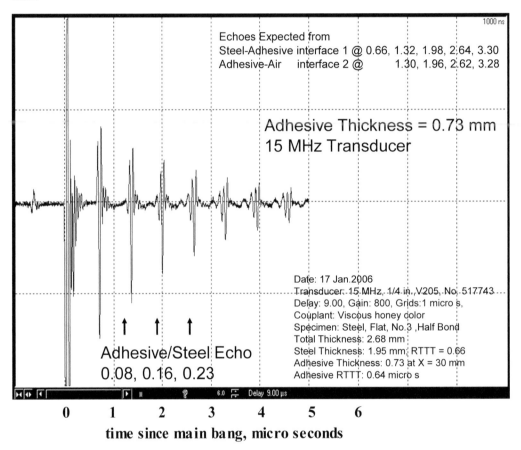

Fig. 5-4(a). Poorly damped oscillations of a typical 15 MHz transducer show how the lack of damping can cause interference with echoes from anomalies that return during the ring-down times just to the right of the reverberating echoes. Delay: 9μs

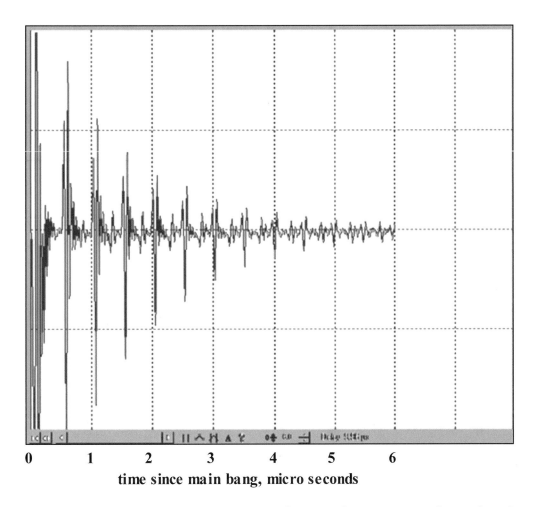

0   1   2   3   4   5   6

time since main bang, micro seconds

Fig. 5-4(b). Poorly damped oscillations of a typical 15 MHz transducer show how the lack of damping can cause interference with echoes from an adhesive layer in a steel bond joint when they return while the transducer is ringing down. The echoes from the bonded adhesive-steel interface 2 appear midway between the larger steel reverberating echoes, 0.23 μs after the first echo from the steel-adhesive interface 1, and at equal time intervals thereafter. Delay before main bang: 9.96μs.

5.2.3 Transducer delay line and acoustic coupling

The transducer is equipped with a plastic delay line to provide ample time for transducer damping (about 6 μs) after the excitation pulse is sent, so that the transducer will have completed its "ring down" of post-pulse vibrations, and thereby have enhanced sensitivity to detect the returning echoes. An example of this ring-down is shown by an A-scan with a typical 15 MHz, and no delay, in Fig. 4-1. Note that the first reverberating echo from the metal-air interface of a 1.92-mm thick steel sheet rings down for 9 cycles, or 0.45 μs.

113

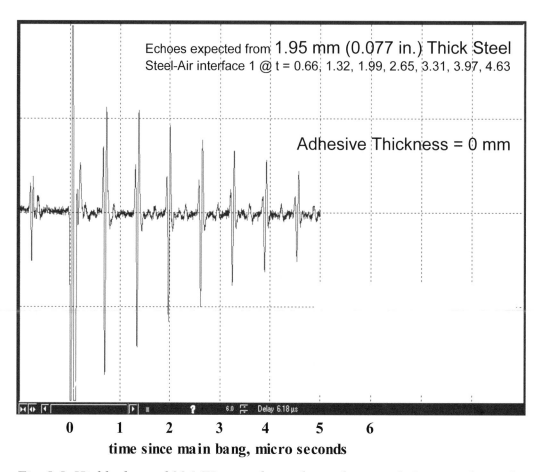

Fig. 5-5. *Highly damped 20 MHz transducer shows the virtual absence of ring down Delay: 6.18 μs*

The A-scan in Fig. 4-1 also shows the residual bias that is concomitant with no delay. The bias diminishes with time, to nearly zero after 6 μs. Therefore the delay avoids that 0.45 μs ring-down time period, and the 6 μs bias-down time period after transducer excitation. Subsequent echoes reverberate in the top metal layer of the joint structure because a significant fraction of the pulse is reflected at the metal-adhesive interface 1, while a complimentary fraction is transmitted through it. Similar pulse transmissions and echo reflections occur at each interface of the bond joint at which echoes arise, and from which they return to be detected by the sending transducer. Reflections from the bonded adhesive-metal interface 2 are identified by arrows in the A-scan of a fully bonded lap joint shown in Fig. 2-9. An example of the echo that returns from the interface between the plastic delay line and the liquid couplant can be seen at -0.7 μs on the left of the A-scan of an adhesively bonded specimen shown Fig. 2-9 and Fig. 5-6.

The plastic delay line was 7.12 mm long, which provided a time delay of 3.1 μs between the occurrence of the "main-bang" excitation pulse at the transducer face and its occurrence at the end of the delay line. Thus the main bang echo is delayed by 6.2 μs, along with all other subsequent echoes from the specimen. This delay is a well-recognized necessity in the pulse-echo NDE of thin metal specimens, where the acoustic velocity is sufficiently high, and the transit times short, in order to provide the transducer sufficient time to damp out resonating oscillations that may be ringing down subsequent to delivering the "main-bang" excitation pulse. Obviously, a quieted transducer can be a more sensitive sensor of the returning echoes. It should be noted here that the ring-down time of the transducers used in this investigation was 0.3 micro s and the bias concomitant with the gain values used disappeared at 6.5 micro s.

A high-density plastic collar, with its 3/4-millimeter tripod protrusions, was fitted onto the specimen-contacting end of the brass cylinder holding the delay line to the transducer, as shown schematically in Fig. 3-1. This tripod fixture was designed to help inspectors maintain the narrow transducer and delay line normal to the specimen surface, in order to facilitate the manual NDE of parts with curved or irregular surfaces. The lift-off resulting from the tripod protrusions created a small 5-mm³ volume between the end of the delay line and the specimen. This volume was filled with couplant which added another 0.84 micro s echo delay to the 6.2 micro s provided by the delay line, thus giving a 7-micro s echo delay.

5.2.4 Acoustic wave train excitation and mode conversion

When the longitudinal ultrasonic wave train is excited by the transducer and the pulse sent from the wear plate on the transducer face through the attached delay line, then through the couplant, into the workpiece or specimen, mode conversion is one of the unintended consequences concomitant with this transmission. Spurious reflection from unavoidable interface surfaces is another. Several sources of mode conversion and spurious reflection are examined here to raise awareness to the issue and point out this source of spurious signals.

Transducers are carefully designed to minimize or eliminate internal reflections from acoustic waves excited when its piezoelectric element expands and contracts in response to the nearly 300-V excitation potential. Longitudinal waves can travel along the intended path trough the wear plate and into the workpiece. They can also travel to the backing plate and experience a minimal reflection that will send low-amplitude secondary pulses toward the workpiece or specimen. Moreover, reverberation reflections in the delay line can be a plausible source of delayed signals reaching the workpiece. This appears to be a likely explanation for some of the regular repeated small spurious reflections that appear in the A-scans at regular, pulse-regulated increments in time.

Beam divergence can cause longitudinal waves to be reflected from the cylindrical walls of the delay line, and there stimulate slower-moving, transversely propagated shear waves and/or surface waves that follow the longitudinal waves to the part where they reconvert from shear waves to late-arriving longitudinal waves. Echoes that may be attributed to this phenomenon can be seen in A-scans such as that shown in Fig. 5-5, appearing midway between the second and third reverberation and in a similar position between subsequent reverberating echoes. The amplitudes of these spurious reflections appear to increase as if by constructive interference, as discussed in chapter 4, in the case of secondary and subsequent reflected echoes from the adhesive layer. Reflections from the sides of the delay line can also arrive late as longitudinal waves, because they took a longer path. These can likewise contribute to spurious anomalous reflections.

Mode conversions can also occur within the part, because hand-held transducers are not likely to be always held in a perfectly perpendicular orientation to the planes in the specimen. In each case proposed here, these causes of anomalous reflections will be late arriving, and in each case examined in the data acquired, except one, these anomalous echoes are always delayed. The important issue here is that they have virtually no impact on the echo signal analysis required to perform the evaluation of the adhesive bond joints.

5.2.5 Acoustic wave propagation and reflection

The physical mechanisms by which the ultrasonic pulse is transmitted, attenuated and reflected in the bond joint has been addressed in chapter 3 on theory, chapter 4 on modeling and again here in opening sections of chapter 5. The explanations of these physical phenomena offered in those preceding chapters form the basis upon which the data will be acquired and interpreted in this chapter.

5.3 Bond-joint specimens

5.3.1 Steel specimens

Most of the bond joint specimens used in this investigation were made from a 1020 steel alloy, with sheet metal thicknesses of 0,73 mm, 1 mm, 1.52 mm and 1.96 mm. These were bonded with a Beta Mate® adhesive, a structural adhesive that is popular in automotive body assembly applications. A more through description of the adhesive is given in chapter six.

5.3.2 Aluminum specimens

The use of aluminum in aircraft and automotive transportation vehicles makes it an essential part of any investigation to develop nondestructive adhesive bond evaluation technology; therefore aluminum bond joints were evaluated.

5.3.3 Polymer composites and plastics

The use of polymer composite and plastic assemblies in automotive and aerospace vehicle applications makes this wide range of materials an essential part of any investigation to develop nondestructive adhesive bond evaluation technology; therefore polymer composite and plastic bond joints were evaluated. However, the high-frequency pulse-echo approach has fundamental limitations for the NDE of adhesive bonds in assemblies made from polymers that generally have acoustic impedances that are very close to those of the polymeric adhesive layer. This limitation was addressed in chapter 4, where modeling showed that another approach is more effective and preferred. That asymmetric Lamb wave technique will be presented in chapter 7.

5.4 Bond-joint scanning, data acquisition and interpretation

Ultrasonic pulse-echo data are acquired and displayed by three convenient methodologies that lend themselves to user-friendly interpretation. The A-scan, B-scan and C-scan terms that identify three common scanning and data presentation methods in ultrasonic NDE were adopted from radar (radio detection and ranging) nomenclature and have a long history of used throughout the NDE community.

When the transducer is held stationary, while the resulting reverberating echoes are displayed in the time domain, the display is referred to as an A-scan. A-scan presentations have been used to present the data in much of this manuscript. When the transducer is moved linearly and the resulting ultrasonic echo data is presented as contrasting bands to show high and low reflection amplitudes at each sequential location, the term B-scan is used. The B-scan represents an acoustic slice through the specimen, along a plane parallel to and coincident with the beam path. The C-scan is performed by moving the transducer in a raster-scan serpentine pattern over the surface over the specimen and the resulting ultrasonic echo data is presented as contrasting pixels in a plane view. The C-scan presents an acoustic image of any designated plane within the specimen that is perpendicular to the beam path. In fact, the transducer must be oriented in such a way as to be able to receive the echo from the pulse sent in all three modes. Data will be presented by each of these modes in this chapter, and that presentation will provide an example of that scan mode.

5.4.1 A-Scans

The reverberating echoes shown in A-scans from unbonded sheet-metal specimens were discussed in chapter 3 on theory, and explained by the schematic illustration in Fig. 3-4 and the subsequent equations derived from it. Figure 3-5(b) shows the predicted exponential decay and associated successive echo amplitude ratio characteristic of that metal, its thickness (1 mm-thick steel) and the reflection coefficients at each surface, here designated as interfaces 0 and 1. Equation (3-24) shows the relationship between the successive echo amplitude ratios in the echo reverberation sequence, and the exponential decay determined by the exponent in equation (3-24). The exponent contains the attenuation coefficient, a, and the thickness, h. The exponential decay shown for this 1 mm thick steel specimen is $A_n/A_{n-1} = 0.88$, characteristic of this thickness of an unbonded 1020 steel specimen and couplant used. When the 1 mm thick steel is bonded to a layer of the cured structural adhesive used in this study, the attenuation is increased, and the ratio drops to 0.76. For 1.95 mm thick unbonded 1020 steel, the ratio is 0.84, but is reduced to 0.73 for the bonded case.

These ratios were virtually constant for each 1020 steel thickness and bond state(bonded or unbonded), over the range of steel thickness and bond states investigated, due to the nearly constant steel attenuation coefficient in the exponent and the constant R values at the air, adhesive and couplant interfaces. Because each constant set of $A_n/A_{(n-1)}$ ratios is concomitant with a specific exponential decay curve, expressed in equation (3-24), these ratios are precise, with coefficients of variation generally ranging from 1% to 6%, but clustering around 3% to 5%. The $A_n/A_{(n-1)}$ data can be seen in table 5.1 for several different steel specimen thicknesses. Thus the successive echo amplitude ratio, $A_n/A_{(n-1)}$, is a reliable and convenient measure of exponential decay rate, or attenuation, and attenuation is a reliable indicator of the state of adhesion at interface 1. Note that only a small portion of this attenuation is caused by the actual attenuation within the material, but caused by the reduced reflection coefficient at the metal-adhesive interface 2, compared to that of a metal-air interface.

Table 5.1 Successive Echo Amplitude Ratios for Unbonded Steel

N of Echoes	Gain	Specm ID No Adh	Thkns of Steel mm	Mean Echo Amplitude Ratio	Median Echo Amplitude Ratio	Coeff. of Vari %
20	1250	Door B2	0.73	0.891	0.903	5.5
19	1200	Door B2	0.73	0.888	0.905	5.0
19	1200	Door B2	0.73	0.879	0.881	6.3
9	1350	SWSB	0.80	0.851	0.861	4.8
10	1350	SWSB	0.80	0.842	0.843	5.0
14	1000	Curv2	0.95	0.884	0.868	5.9
8	1200	SWBrn/Slv	1.52	0.808/.873	0.808/.875	3.2/1.6
8	1350	SWSB	1.77	0.818	0.819	3.1
8	1350	SWSB	1.77	0.825	0.831	3.9
6	1200	3 HB	1.95	0.838	0.830	3.4
7	1200	3 HB	1.95	0.835	0.846	4.9
8	1350	3 HB	1.95	0.819	0.827	2.9
2	1050	SWSB	5.70	0.698	0.698	4.4
2	1050	SWSB	5.70	0.628	0.628	12.3
1	1350	SWSB	6.32	0.654	0.654	

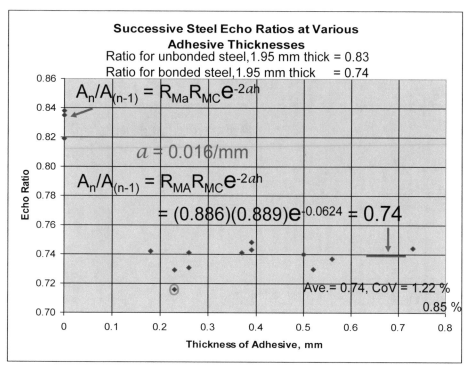

Fig. 5-6 Successive echo amplitude ratio, A_n/A_{n-1}, is constant over all adhesive thicknesses to illustrate that the reflection coefficient, not the adhesive thickness, increases echo attenuation. Here A_n/A_{n-1} drops from 0.83 to 0.74 because of the presence of the adhesive bonded to the steel at interface 1.

The presence of bonds at the first interface, between the first metal layer and the adhesive, are recognized by the increased attenuation rate of echoes reverberating in the first metal sheet, and by echoes from the second adhesive interface, as illustrated by the contrasting examples shown in Fig. 5-5 and Fig. 5-7.

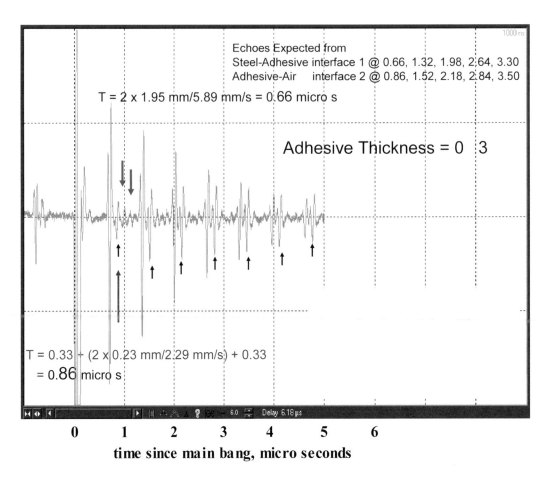

Echoes Expected from
Steel-Adhesive interface 1 @ 0.66, 1.32, 1.98, 2.64, 3.30
Adhesive-Air   interface 2 @ 0.86, 1.52, 2.18, 2.84, 3.50

T = 2 x 1.95 mm/5.89 mm/s = 0.66 micro s

Adhesive Thickness = 0.3

T = 0.33 + (2 x 0.23 mm/2.29 mm/s) + 0.33
= 0.86 micro s

Delay 6.18 μs

0   1   2   3   4   5   6

time since main bang, micro seconds

Fig. 5-7. A-scan of 1.95 mm thick half-bonded 1020 steel sheet, bonded to a 0.23 mm thick adhesive layer, showing echoes from the adhesive-air interface 2, marked by lower arrows, at 0.86 µs, 1.52 µs, 2.18 µs, 2.84 µs, 3.50 µs, etc. Faint echoes marked by upper arrows are from highly attenuated reverberations in the adhesive. Note the increased attenuation in metal echo reverberations from An/A(n-1) = 0.83 in Fig. 5-5 to 0.74 here. 20 MHz transducer Delay: 6.18 µs

A review of Fig. 5-7 reveals the following for each interface echo sequence:

• Plastic delay line-to-couplant echo should show an initial small up-swing, followed by a major negative excursion because R < 0

• Couplant-to-metal (main bang) echo should show an initial down-swing, followed by a major positive excursion because R > 0

• Metal-to-adhesive reverberating echoes should show an initial small up-swing, followed by a major negative excursion because R < 0

• Adhesive-air interface echoes (multiple echoes resulting from reverberations in the metal layer) should show an initial small up-swing, followed by a major negative excursion because R < 0

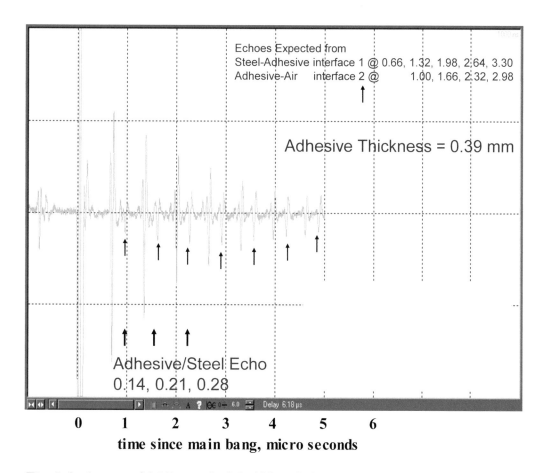

Fig. 5-8. *A-scan of 1.95 mm thick half-bonded 1020 steel sheet, bonded to a 0.39 mm thick adhesive layer, showing echoes from the adhesive-air interface 2, marked by lower arrows, at 0.96 µs, 1.66 µs, 2.32 µs, 2.98 µs, 3.60 µs, etc.*
20 MHz transducer Delay: 6.18 µs

The A-scan shown in Fig. 5-8 clearly shows the negative phase preference of echoes returning from the adhesive-air interface concomitant with an unbond at interface 2, as predicted by theory in chapter 3 and modeling in chapter 4. This is the distinguishing characteristic of acoustic reflections from an unbonded adhesive layer, with a smooth surface that is parallel to the metal surface upon which the transducer is placed. The successive amplitude echo ratio is more attenuated than when interface 2 is unbonded, as expressed in equation (3-24), because R_{Ma} is replaced by R_{MA}.

$$A_n/A_{n-1} = R_{Ma}R_{MC}e^{(-2ah_M)} \qquad (3-24)$$

where

A_n is amplitude of nth echo exiting the reverberation,

A_{n-1} is amplitude of the previous echo,

R_{Ma} is reflectance at the metal-air interface 1,

R_{MC} is reflectance at metal-couplant interface 0,

a_M is the attenuation coefficient of the metal for the frequency distribution characteristic of the 20 MHz pulse, and

h_M is the thickness of the metal specimen layer interrogated.

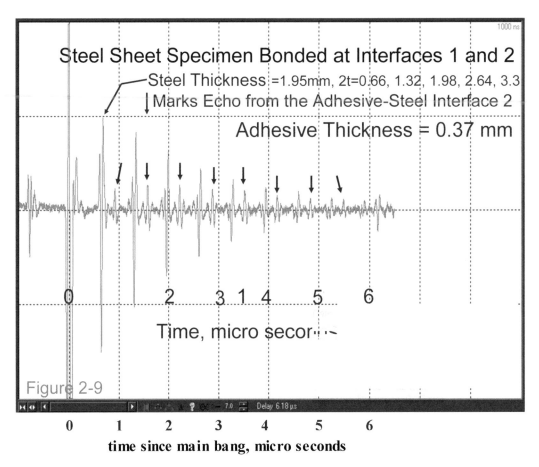

Fig. 2-9. A-scan of 1.95 mm thick full-bonded 1020 steel sheets, bonded with a 0.37 mm thick adhesive layer, showing echoes from the adhesive-metal interface 2, marked by upper arrows, with times of flight, 2t, in μs. 20 MHz transducer Delay: 6.18 μs

An example of A-scans from fully bonded metal specimens is shown in Fig. 2-9 and Fig. 5-9. Note the positive phase enhancement characteristic of echoes reflected from a metal-adhesive interface, as predicted by equation (3), where $R_{AM} > 0$. This echo pattern identifies the adhesive bond as being adhered to the metal at interface 2, as well as at interface 1. Obviously, the acoustic energy would not reach interface 2 without adhesion at interface 1.

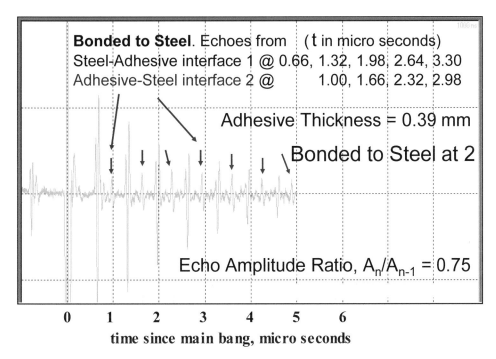

*Figure 5-9. A-scan showing decay of successive reverberations in steel sheet bonded to another sheet by an adhesive layer. Note positive enhancement of the phase of echoes reflected from the **bonded** adhesive-steel interface 2.*

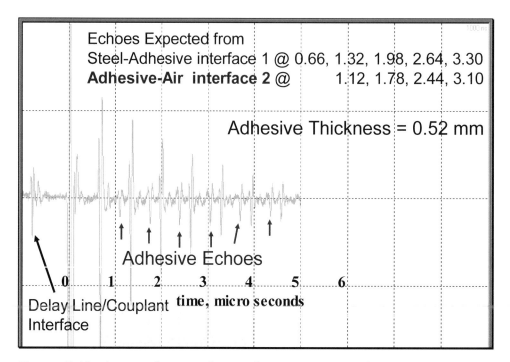

Figure 5-10. A-scan showing decay of successive reverberations in steel plate bonded only to an adhesive layer. Note phase of adhesive-air echoes.

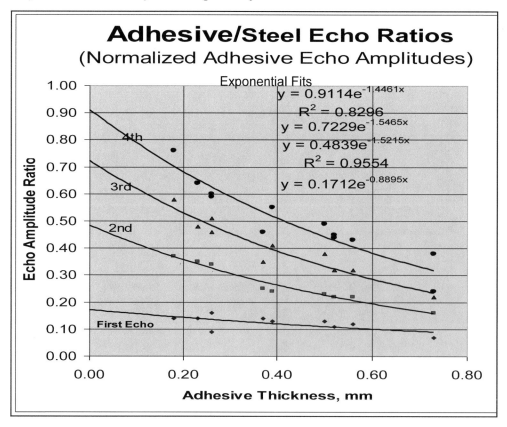

Fig. 5-11 Ratios of the amplitude of echoes returning from the adhesive-air interface 2 to the corresponding echoes returning from the metal-adhesive interface 1

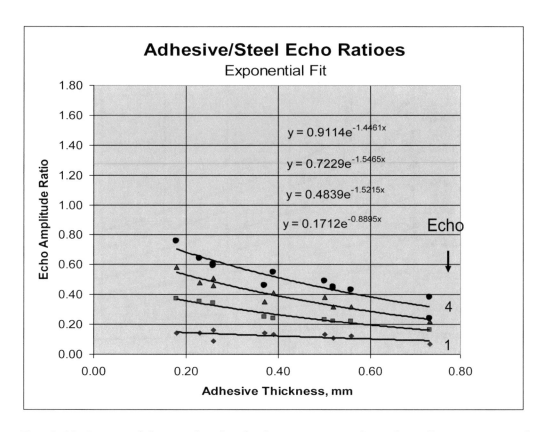

Fig. 5-12 Ratios of the amplitude of echoes returning from the adhesive-air interface 2 to the corresponding echoes returning from the metal-adhesive interface 1. (Repeated with compressed scale for ease of comparison to Fig. 4-14)

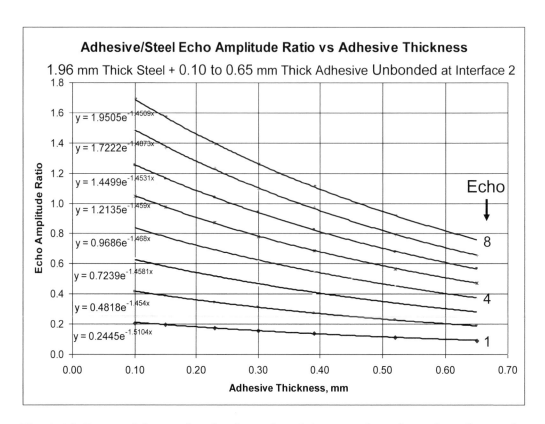

Fig. 4-14 *Ratio of the amplitude of simulated A-scan echo n from the adhesive layer at interface 2, to the amplitude of the corresponding simulated echo n from the steel sheet at interface 1. Comparing this family of curves with data validates the model.* (Repeated here for convenience)

Interferences in completely bonded joints can be observed. The positive phase of A-scan echoes reflected from the adhesive-metal interface, interface 2, of fully bonded joints, have their amplitudes reduced to about 2/3 of what they would have been without the interference from negative-phase echoes reflected from the metal-air interface. This interference occurs when the second steel sheet has the same thickness as the top sheet, or half-multiples thereof, but is insignificant in this investigation, because it does not affect the outcome of the bond-state analysis. Therefore this interference was not considered in the modeling of chapter 4. The schematic diagram of a bond joint shown in Fig. 5-13 shows the coincidental arrival times at 3, of the negative-phase echo from the metal-air interface and the positive-phase echo from the adhesive-metal interface. When the first three adhesive-to-metal echo amplitude ratios are taken for the bonded joints, and divided by the corresponding ratios from the unbonded joints, the results are 0.64, 0.70 and 0.65, respectively. This 33% reduction is attributed to the interference.

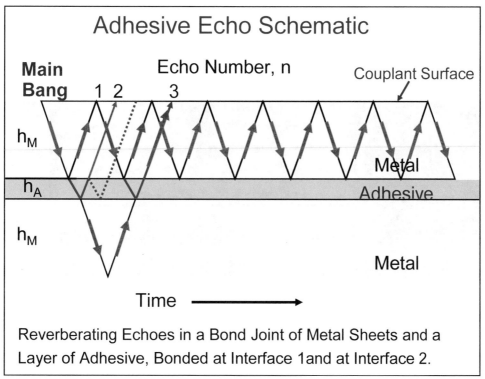

Fig. 5-13 Illustration of interference of echo from metal-air interface 3, arriving at time 3, the same time as the echo from the adhesive-metal interface 2, when thickness of top and bottom metal sheets are equal.

5.4.2 B-Scans

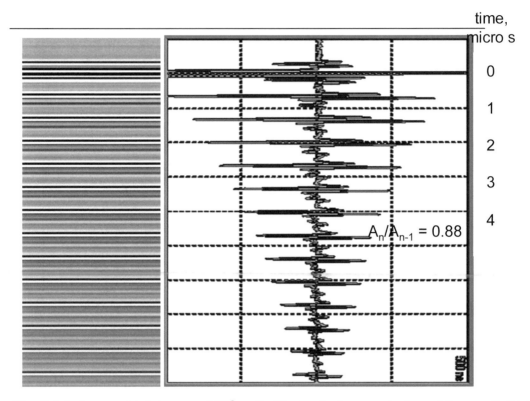

Fig. 5-14 A-scan (right) rotated 90° to facilitate the interpretation of B-scan (left). The B-scan presentation was acquired as an A-scan, by not moving the transducer.

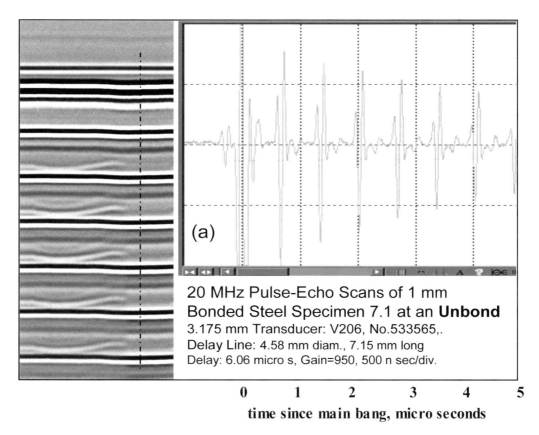

20 MHz Pulse-Echo Scans of 1 mm
Bonded Steel Specimen 7.1 at an **Unbond**
3.175 mm Transducer: V206, No.533565,.
Delay Line: 4.58 mm diam., 7.15 mm long
Delay: 6.06 micro s, Gain=950, 500 n sec/div.

time since main bang, micro seconds

Fig. 5-15(a) B-scan (left) with companion A-scan (right) acquired at the unbonded position identified by the vertical line in the B-scan, so $A_n/A_{(n-1)} = 0.88$

The absence of a series of echoes from the adhesive layer seen in the B-scan of Fig. 5-14(b), illustrates the importance of being able to detect the presence of adhesion at interface 1 by increased attenuation of the echoes reverberating in the metal. The non-parallel orientation of the adhesive layer can be seen in the B-scan, and is the cause of no reflections from it reaching the transducer which was located at the top of the B-scan image, centered along the axis coincident with the vertical broken line.

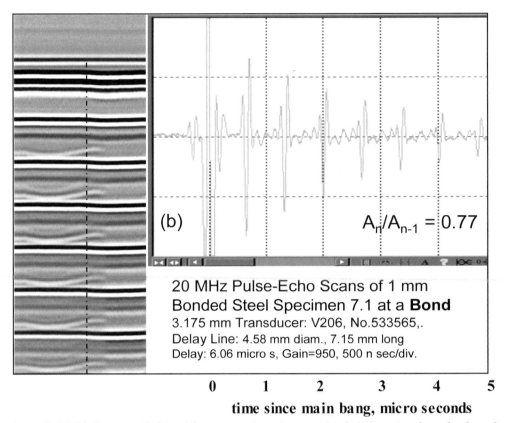

(b)

$A_n/A_{n-1} = 0.77$

20 MHz Pulse-Echo Scans of 1 mm
Bonded Steel Specimen 7.1 at a **Bond**
3.175 mm Transducer: V206, No.533565,.
Delay Line: 4.58 mm diam., 7.15 mm long
Delay: 6.06 micro s, Gain=950, 500 n sec/div.

0   1   2   3   4   5

time since main bang, micro seconds

Fig. 5-15(b) B-scan (left) with companion A-scan (right) acquired at the bonded position identified by the vertical line in the B-scan, so $A_n/A_{(n-1)} = 0.77$, showing higher echo attenuation due to adhesion at interface 1. The absence of a series of echoes from the adhesive layer at interface 2 indicates that the reflections from that non-parallel adhesive layer do not reach the transducer because of the orientation of the layer.

5.4.3 C-Scans and acoustic microscopy data

Ultrasonic pulse-echo C-scan presentations were made using a standard 50 MHz spherically focused transducer to evaluate a collection of adhesively bonded steel specimens, with substrates that ranged in thickness from 1 mm to 2 mm. The SAM provides images at different depths into the substrate or the adhesive layer, depending on where the focus of the transducer is directed. More details are provided in chapter 2 on theory, in section 2.3, and section 2.5.

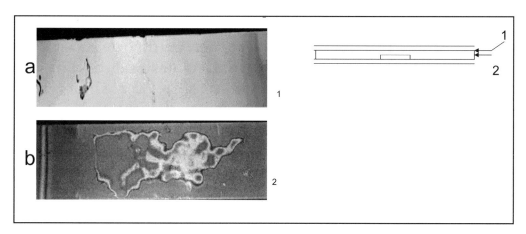

*Fig. 5-16(a) and (b) Scanning acoustic microscope C-scan presentation acquired on an adhesive bond joint made from 1 mm-thick steel sheets, using a 50 MHz spherically focused transducer. Scan **a** is the image acquired when the transducer was focused to acquire echoes from the plane at interface 1, as indicated in the schematic of the cross-section shown at upper right. Scan **b** is the image acquired when the transducer was focused to acquire echoes from the adhesive layer near the plane at interface 2. (I. Severina)*

In Fig. 5-16(a), the transducer is focused to acquire pulse-echo data from the interface 1 plane. In Fig. 5-16(b), the transducer is focused to acquire pulse-echo data from near the interface 2 plane. Note that the unbonded region near the center of Fig. 5-16(b) appeared mostly bonded at interface 1, shown in Fig. 5-16(a), but had become an unbonded region about 40 mm long at interface 2 and can be clearly seen near the center of the scan image.

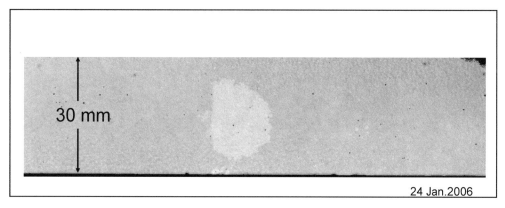

Fig. 5-16(c) Scanning acoustic microscope (SAM) C-scan presentation acquired by using a 50 MHz spherically focused transducer on a bonded steel specimen. Here the transducer is focused to acquire pulse-echo data from the interface 1 plane. Note that the unbonded region near the center. (I. Severina)

C-scans such as these were made with high-resolution SAM on many of the adhesively bonded specimens used in this investigation, and therefore they provided a reliable, high-resolution set of reference specimens on which to evaluate the effectiveness and reliability of this new phase-sensitive ultrasonic technique. These scans, as in the examples shown in Figs. 5-16(a), 16(b) and 5-16(c), very valuable in providing clearly identified locations on bonded specimens that were unbonded only at one interface with a kissing unbond, usually where the entire joint was filled with adhesive, or they were cases in which metal "spring-back" or adhesive post-cure shrinkage caused disbanding in the presence of adhesive.

It was very important to have realistic unbonds such as these, and to have available these confirmed **kissing unbonds** that were **not created artificially** to evaluate the effectiveness and reliability of the method. The 20 MHz method agreed with the SAM results in each case, to the extent of its 3mm to 4mm resolution capability.

5.5 Attenuation determinations

In order to evaluate the validity of the model developed in chapter, it was necessary to obtain realistic values for the attenuation of the materials used in this investigation. These values could be calculated from each datum by equation (3-26), or by a more statistically reliable method of taking half the coefficient of the exponent in the exponential curve that fits the $A_n/A_{(n-1)}$ data plotted in Fig. 3-6, and similar treatments, in which the thickness of the material was varied throughout a sufficient range, with the $A_n/A_{(n-1)}$ value at 0 thickness being the product of the two reflection coefficients for the interfaces bordering the material.

5.5.1 Attenuation determination in sheets and blocks

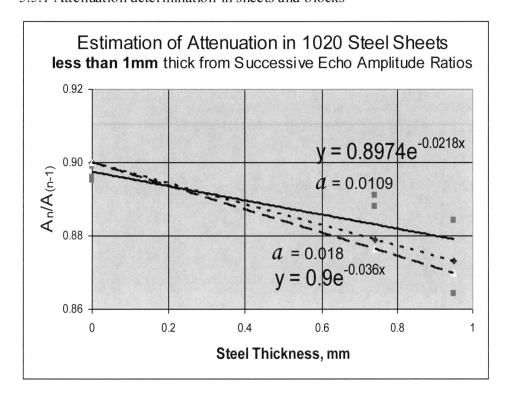

Fig. 3-6 (Repeated for convenience) Successive echo amplitude ratios, $A_n/A_{(n-1)}$, plotted as a function of the thickness of 1020 sheet steel, for the determination of ultrasonic attenuation in the material.

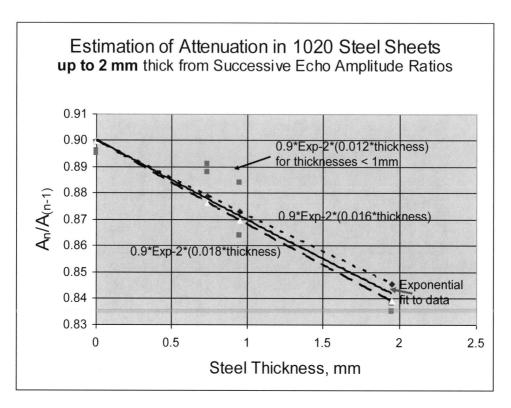

Fig. 5-17 Successive echo amplitude ratios, $A_n/A_{(n-1)}$, plotted as a function of the thickness of 1020 sheet steel, for the determination of ultrasonic attenuation in the material.

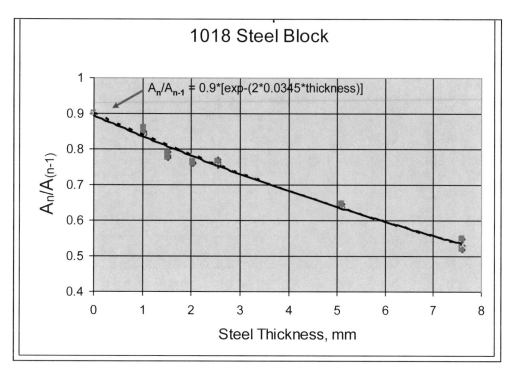

Fig. 5-18 Successive echo amplitude ratios, $A_n/A_{(n-1)}$, plotted as a function of the thickness of 1018 steel thickness-gauge blocks, for the determination of ultrasonic attenuation in the material.

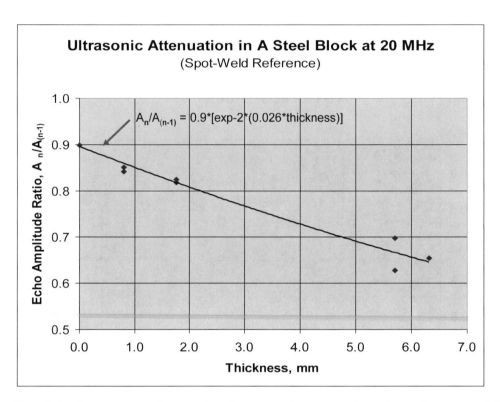

Fig. 5-19 Successive echo amplitude ratios, $A_n/A_{(n-1)}$, plotted as a function of the thickness of a steel spot-weld reference block, for the determination of ultrasonic attenuation in the material.

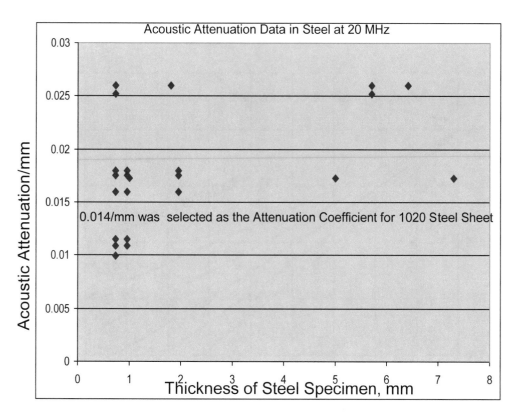

Fig. 5-20 Summary of acoustic attenuation measured in three groups of steel specimens. Note the separation of the attenuation data into three statistically significant groups, independent of thickness. The lowest (0.010/mm to 0.012/mm) and mid-range values (0.016/mm to 0.018/mm) were measured in 1020 steel sheet specimens. The highest range group shown (0.025/mm to 0.026/mm) was measured in a steel spot-weld reference block. Not shown are the even higher values (0.034/mm) obtained from two 1018 steel thickness-gauge calibration step blocks.

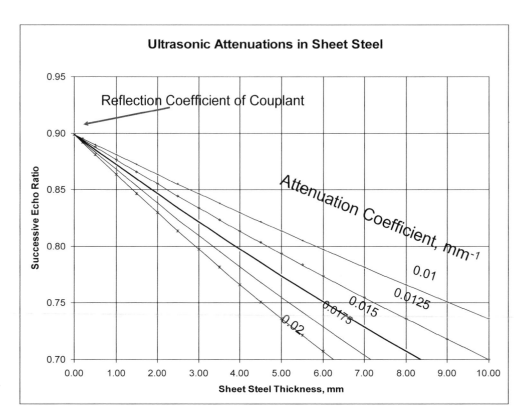

Fig. 5-21 Calculated family of curves that show the relationship among various values for attenuation coefficients, reflection coefficients, thickness and successive echo amplitude ratios, An/A(n-1), for echoes reverberating in 1020 steel bounded by couplant and air, for a wide range of steel thicknesses

5.5.2 Attenuation determination in adhesive layers

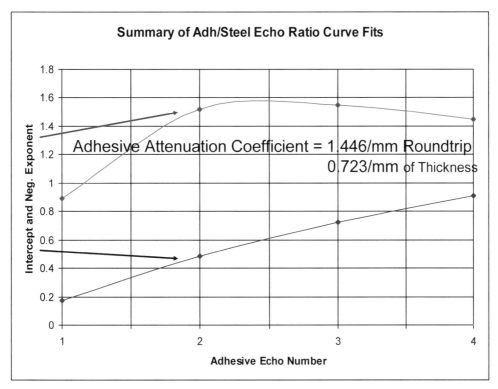

Summary of Adh/Steel Echo Ratio Curve Fits

Adhesive Attenuation Coefficient = 1.446/mm Roundtrip
0.723/mm of Thickness

Fig. 5-22 Cross-plot summaries of adhesive-to-steel echo ratio data plotted in Fig. 5-11 to provide an approximation of the attenuation coefficient for the adhesive layer.

5.5.3 Attenuation determinations during NDE of bond joints

The ultrasonic pulse-echo technique provides data which can be used to determine the attenuation of the metal substrate sheets that are bonded. This data is provided without additional cost of instrumentation, effort or inspection operation. It is valuable additional information in that it provides concurrent monitoring of changes in the metallurgical history of the metal used to fabricate the bonded component. Unintended or unsuspected changes in the metallurgical history of the metal supplied to the forming operation can cause serious problems, when these changes are introduced after the forming parameters have been established for a specified metal alloy that has a specific metallurgical history. This technique can identify such changes by detecting changes in attenuation.

Figure 5-20 shows a summary of acoustic attenuation acquired on what appears to be three groups of different steel specimens. Note the separation of the attenuation data into three statistically significant groups, independent of thickness. The lowest (0.010/mm to 0.012/mm) and mid-range values (0.016/mm to 0.018/mm) were measured in 1020 steel sheet specimens. The highest range group shown (0.025/mm to

0.026/mm) was measured in a steel spot-weld reference block, known to be different from the sheet metal specimens. Not shown are the even higher values (0.034/mm) obtained from two 1018 steel thickness-gauge calibration step blocks. These values were derived from the data plotted in Fig. 5-18.

5.6 Confirmation of NDE results by destructive tests

The effectiveness of NDE results is validated by mechanical testing for the property of interest, or by actual performance evaluation of components in simulated or actual service.

5.6.1 Bonded metal components

Laboratory feasibility has been demonstrated on bonded test specimens where the NDE results were confirmed by destructive tests and/or by reliable, high-resolution techniques, such as SAM, as discussed in this chapter. Moreover, manufacturing feasibility has been demonstrated by several adhesive bond NDE operations on automotive vehicle bodies in assembly plants. In each case, the validity of the NDE method was confirmed with no false negatives. The converse can be correctly stated for false positives, but conveys little meaning, because bond regions suspected of being bonded were generally not destructively tested.

5.6.2 Bonded plastic and polymer composite specimens and components

Since the NDE of these specimens is not recommended by this technique, this issue will be addresses extensively in chapter 7 where bonds made from these materials are evaluated.

5.7 Advantages and Limitations of the Method

5.7.1 Advantages on aluminum and steel assemblies

The pulse-echo technique is very effective evaluating adhesion in metal joints joined by polymer adhesives, because the acoustic impedance mismatch is high, resulting in a strong echo from the interface.

5.7.2 Limitations on polymer composite and plastic assemblies

Since this NDE technique is accomplished by the acquisition and analysis of acoustic echoes that return from the ultrasonic interrogation of bond joints that have interfaces between layers with large acoustical impedance mismatch, it is not very effective where this difference does not exist, such as when plastics and polymer composites are bonded with a polymer adhesive.

5.8 Discussion of results and analysis of errors

The need to inspect adhesive bonds in mass-produced vehicles increases as the demand for improvements in fuel-efficiency, performance, corrosion resistance, body stiffness and style increases, because the use of adhesives for structural joining continues to grow in order to help meet those demands by providing greater design flexibility, more materials choices and wider process options. Although the growth of adhesive bonding is increasing, the history of adhesive bond performance in vehicle body structures [1] indicates the need for a robust method of assuring adhesive bond integrity during the manufacturing process.

Extensive research has been done to development NDE technologies to meet this growing need [1, 2, and 3] in the aerospace and automotive industries. Each research and development effort has generally focused on the needs and requirements specific to the application identified by the industry driving the research, and each effort has provided progress in the journey to meet those needs. The research and develop effort reported here is another step in the journey to provide a robust, effective nondestructive adhesive bond inspection method, with high resolution, that requires access from only one side, uses a small probe that can interrogate limited-access locations, and has the potential for a wide variety of applications in automotive manufacturing.

5.9 Conclusions and next steps

Laboratory NDE of prepared specimens and automotive body samples from production operations show that joint integrity at both interfaces can be robustly evaluated using a 20 MHz, 3-mm transducer element, with a 6-mm diameter, 7-mm long standard delay line with a couplant.

The indications and the resulting interpretations show how the presence of bonds at the first interface, between the first metal layer and the adhesive, are recognized by the increased attenuation rate of echoes reverberating in the first metal sheet, and by echoes from the second adhesive interface, when they exist. Bond integrity at the second interface is evaluated by a phase-sensitive analysis of the echoes reflected from that adhesive-metal interface.

Attenuation coefficients were also determined for the various materials used.

Attenuation coefficient, $a = 0.00998$/mm on 0.73 mm thick steel door stock

Attenuation coefficient, $a = 0.016$/mm on various steel specimen thicknesses

Attenuation coefficient, $a = 0.0252$/mm on steel step block

Attenuation coefficient, $a = 0.026$/mm on curve-fit to steel step block

Reflection coefficient at steel-couplant interface, $R_{Mc} = -0.89$

When adhesive is bonded to the metal substrate at interface 1, the combined reflection coefficients change from the product of those for air and couplant, $R_{Ma}R_{Mc}$, to the product of those for adhesive and couplant, $R_{MA}R_{Mc}$, and because the reflectance for a metal-air interface (very nearly 1) is greater than the reflectance for a metal-adhesive interface, $R_{Ma} > R_{MA}$, the results is a significant increase in the exponential decay of the reverberating echoes and a concomitant reduction in A_n/A_{n-1}, as predicted by equation (3-24). A steel specimen with identical composition, metallurgy and geometry, but bonded to an adhesive layer, yields an $A_n/A_{n-1} = 0.77$, whereas the unbonded ratio is 0.88. This difference of 0.11 is very statistically significant, because it represents a difference of 3.6 standard deviations from the 0.88 value observed for the unbonded counterpart. This highly significant difference is supported by precise coefficients of variation for these ratios that range from 2% to 6% for the data acquired during this study. These statistics infer a level of confidence greater than 98% that these two values, separated by 3.6 standard deviations, belong to different populations.

The conclusion can therefore be drawn that measuring the exponential decay rate of the reverberating echoes by computing A_n/A_{n-1} for several successive echoes is a reliable indicator of the bond state at interface 1. The bond state at interface 1 can be reliably determined by this technique even when there is no echo from the boundary of the adhesive layer at interface 2. The absence of such an echo could be caused by scattering from a rough, non-reflecting adhesive surface at 2. When a smooth reflecting adhesive surface does exist at interface 2, but there is no bond to the metal substrate there, a predictable echo pattern will appear to indicate this bond state.

Another technical approach is recommended in chapter 7 for bond joints assembled from materials in which these significantly large differences in acoustic impedance do not exist between the substrate material and the adhesive material. Bonded assemblies fabricated from plastics and polymer composites are among them, because most automotive structural adhesives are polymers with acoustic impedances comparable to those of other polymeric materials.

Improvements in the inspection and data analysis techniques are underway as demonstrations of manufacturing feasibility proceed in several vehicle body assembly applications. Many of these manufacturing feasibility demonstrations have proven successful and applications implemented.

6. Ultrasonic Analysis of the Degree of Cure and Cohesion within the Adhesive Layer of the Bond Joint

6.1 Why evaluate the state of adhesive cure

The often-cited advantages of adhesive bonding depend on the precise control of critical parameters which influence the performance characteristics of the adhesive, such as its adhesion to the substrate, its cohesive strength, its toughness and durability. So consequently, the overall joint performance depends upon the precise control of these critical parameters, 16 of which were listed in chapter 1. Prominent among them are the thermal and environmental histories of the adhesive, because these are primary among the critical parameters which have an influence on the degree of cure, or over cure, of the adhesive in the joint, and therefore the bond joint performance. It is well-known that an insufficient cure state for thermoset adhesives can be caused by both improper cure temperature and time, and the cure degree of the adhesive correlates with both the cohesive strength of the adhesive material and the adhesion strength; hence improper curing can lead to joint failure by both cohesion and adhesion failure modes.

Network formation, commonly called crosslinking, of the thermoset adhesive is a complex process in which three-dimensional linkages are formed throughout the polymer during the curing process. While curing, the thermoset adhesive undergoes significant changes from a low-molecular weight liquid or paste, to a highly cross-linked network, as its modulus and strength increase to those of a solid. These chemical changes in the epoxy during polymerization result in changes in both the elastic constants and glass transition temperature. The rate and degree of chemical curing also depend on resin-to-hardener ratio, which thus influences the fully developed epoxy cure state, or the extent to which the crosslinking reaction can advance.

Therefore, since proper curing depends on multiple parameters, such as temperature, time, resin-to-catalyst ratio, etc., monitoring and assuring the progress of the curing process is essential to assuring the quality of a bond joint. Effective in-process monitoring requires the monitoring of the development of this epoxy network architecture that plays a significant role in the mechanical behavior of the bond joint. Thus, cure monitoring involves real-time tracking of the varying physical state resulting from chemical reactions occurring during the process, and such in-process monitoring will provide efficient cure temperature-time optimization that will yield joints that exhibit the aesthetic characteristics and structural performance requirements in the assembly.

It is important to note here that the application of NDE techniques to adhesive bonds in production environments is often undertaken before the bonds are cured to the final state by subsequent thermal processes, such as in paint curing ovens. Such situations may not require monitoring the state of cure of the adhesive in the joint, especially in well-established mass-production manufacturing operations, but the understanding derived from this investigation into the physical and mechanical idiosyncrasies of the uncured and partially cured adhesive, and their effect on the propagation and reflection of the ultrasonic signal, can significantly enhance the effectiveness of the application of the ultrasonic NDE techniques, as well as identify the limits of the application's effectiveness.

6.2 Progress in the NDE of the state of adhesive cure

A method of evaluating the degree of cure of the adhesive in a bond joint has been reported by Maeva and her colleagues [187] and provides another important component in the completion of a full suite of NDE methods for the comprehensive evaluation of adhesive bond joints. These components include the detection of (1) adhesion at the first interface, (2) adhesion at the first and second interfaces, and (3) the proper degree of cure of the adhesive. The third component of the set is not now routinely performed while performing the first two, because it requires certain known bond-joint parameters which are acquired during the evaluations that comprise those two previous components of the suite. Moreover, in production applications, manufacturers prefer performing the NDE before the adhesive is cured; therefore making the application of a cure-monitoring NDE technique more likely in the final stages of the manufacturing process to evaluate the effectiveness of the curing process.

Acoustic techniques have been used to monitor cure of thermoset adhesive during cure reactions and to evaluate the mechanical and cohesive properties of the material in the adhesive joint after cure. Longitudinal acoustic velocity and attenuation were used to monitor the elastic moduli of the adhesive and to investigate molecular network development during reaction at different temperatures. Ability of high resolution acoustic microscopy to study the development of adhesive microstructure during cure reactions was also demonstrated. Correlation of acoustic velocity values with cohesive properties of the material and joint strength was demonstrated, indicating that such a method can be used to monitor the development of the adhesive strength during the cure process and to nondestructively detect insufficiently cured joints immediately after the cure stage in manufacturing.

6.3 Optimum method of monitoring the state of adhesive cure

Conventional ultrasonic acoustic inspection is known to be a reliable and effective method for nondestructive defect identification and evaluation of the adhesive joint quality, but the effectiveness of these conventional defect detection methods for evaluation of elastic properties in on-line applications is limited. This limitation is due to the conventional flaw-detection approach to interrogation and interpretation of the propagation of the acoustic waves. In addition to flaw-detection NDE, high frequency ultrasonic methods give indications that provide an assessment of the physico-mechanical properties of the material, such as density, elasticity and viscosity. Because these are the properties that undergo change during the curing process, the ultrasonic interrogation approach has high potential for effectiveness in measuring them in the bond joint.

Recent advances in acoustic microscopy now provide NDE techniques by which evaluation of the state of cure of the adhesive within the bond joint can be performed. Although the conventional acoustic technique is still the most reliable method for non-destructive defect detection, its use for evaluating elastic properties in on-line applications is still quite limited, and since those critical mechanical properties of the adhesive bond joint, such as strength and modulus, are strongly related to the cure state of the adhesive, an NDE technique is needed that will monitor the state of adhesive cure during and after the process. Acoustic techniques have been used by researchers to analyze elastic properties and to study in-process molecular network development [80, 188, 189, 190, and 191]. Although much of the work reported therein was performed in transmission mode, transmission mode is not convenient for industrial applications because industrial applications usually limit inspection access to only one side of the specimen or workpiece. In this current work, however, the reflection mode has been applied to in-process monitoring of the changes in acoustic and elastic properties within the adhesive during the reactions that lead to curing.

The high resolution capability of acoustic microscopy provides an opportunity to nondestructively visualize the microstructure of the material, as well as investigate the distribution of the different physico-mechanical properties throughout the material comprising the structure under interrogation, including the adhesive/substrate interface.

The objectives of the work reported in this chapter are to characterize and understand acoustic properties of the epoxy-based adhesive during cure reaction at elevated temperatures and to demonstrate how the acoustic parameters correlate with cohesive properties of the adhesive. Another goal is to evaluate the ability of high resolution acoustic microscopy for monitoring the development of adhesive microstructure during cure reactions.

6.4 Materials, equipment and monitoring methods

A commercial grade, one-component, structural epoxy adhesive, currently used in many automotive applications (BETAMATE 1496 [R]) was used in this investigation. The manufacturer recommends 30 minutes at 180 C as the optimum curing regime for this adhesive. Temperature dependence of the acoustic properties of the completely cured adhesive was investigated using an ultrasonic polymer characterization system (Industrial Materials Institute, NRC Canada). The transmission mode at 2.5 MHz and pressure of 100 bar was used.

Isothermal monitoring of the adhesive cure was performed at the frequency of 15 MHz using Panametrics[TM] 5073PR pulse-receiver connected to an oscilloscope (Tektronix[TM] TDS 520C). The experimental setup for this cure monitoring procedure is illustrated schematically in Fig. 6-1. The epoxy sample was placed in the specially designed preheated cell. The isothermal conditions were maintained with a temperature controller (Digi-Sence[TM]). The epoxy resins reached the required temperature within 2-3 min. Acoustic parameters were taken at 0.5 min. interval.

The pulse-echo mode was used for ultrasonic excitation and data reception, where the same transducer emits the outgoing pulse and receives the reflected echoes. Longitudinal sound velocity C_L was calculated according the formula:

$$C_L = \frac{2h}{\Delta t} \qquad (6\text{-}1)$$

where
h is the thickness of the specimen and
Δt is the difference between the time of flight of echoes reflected from the upper and lower interfaces.

Thicknesses of these specimens vary between 2.5 mm and 3.0 mm.

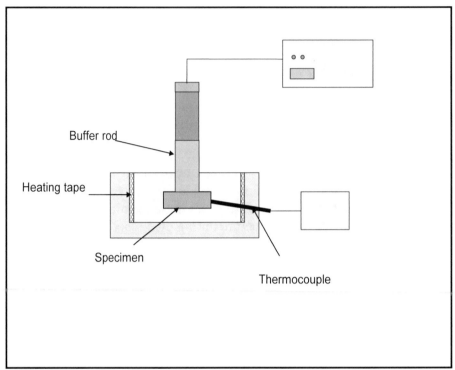

Fig. 6-1. Experimental apparatus for monitoring adhesive cure by an ultrasonic pulse-echo technique.

During isothermal cure, only relative attenuation is monitored and changes in acoustic attenuation $\Delta\alpha$ are calculated according to following:

$$\Delta\alpha = \frac{1}{x}\left[\ln\left(\frac{A_{final}}{A_{cure}}\right)\right] + \ln\left(\frac{T_{12cure}T_{21cure}R_{23cure}}{T_{12final}T_{21final}R_{23final}}\right)$$

(6-2)

where

A_{final} and A_{cure} are the amplitudes of the signal reflected from the lower interface,
A_{final} is amplitude data acquired at the end of the cure reaction
A_{cure} is amplitude data acquired at certain indicated monitoring points during the process.
T is the transmission coefficient at the interface state indicated by the corresponding subscripts
R is the reflection coefficient at the interface state indicated by the corresponding subscripts
subscripts *12* identifies the interface between the epoxy buffer rod and the adhesive specimen
subscripts *23* identifies the interface between the adhesive specimen and the aluminum cell
subscript *cure* identifies characteristics at indicated monitoring points during the process
subscript *final* identifies characteristics at the final cure state of the adhesive

Transmission and reflection coefficients were calculated by the following formulas:

$$T_{ij} = \frac{2Z_i}{Z_i + Z_j}$$

(6-3)

$$R_{ij} = \frac{Z_j - Z_i}{Z_j + Z_i},$$

(6-4)

146

where

Z is the acoustic impedance of materials i or j, where i has values 1 and 2 when j has values 2 and 1, respectively,

$Z = C\rho$, as defined in (2-30) and discussed in subsequent chapters 3, 4 and 5.

It was shown in previous experiments that density increases by 2-4 % during cure for this type of adhesive, so we neglect density variations during cure. Relative attenuation for completely cured epoxy was then brought into accord with the absolute attenuation, α_{final}, value obtained by the acoustic method in transmission mode at corresponding temperature. Thus, absolute attenuation α was calculated as

$$\alpha = \Delta\alpha + \alpha_{final}$$

(6-5).

Experimentally it was shown that the last term in the formula for attenuation (6-2), considering changes in the reflection and transmission coefficients, varies during the reaction from -0.0013 to 0.008. Thus, the contribution of changes in R and T can be neglected during calculation of the attenuation value.

If the sample dimension, normal to the direction of propagation of acoustic longitudinal waves, is much larger then the wavelength, propagation is guided by bulk longitudinal modulus L which relates to bulk, K, and shear, G, modulii as:

$$L = K + \frac{4}{3}G$$

(6-6).

As seen from equation (6-6), the information about both mechanical properties, compression modulus, K, and shear modulus, G, are contained in the longitudinal modulus L. Note that equation (6-6) also agrees with the outcome stated by equation (3-20).

The acoustic attenuation and velocity data were used for calculation of storage L', loss L" bulk longitudinal moduli as well as loss factor tgδ according following formulas [190]:

$$L' = \frac{\rho C_L^2 \left[1 - \left(\frac{\alpha\lambda}{2\pi}\right)^2\right]}{\left[1 + \left(\frac{\alpha\lambda}{2\pi}\right)^2\right]^2}$$

(6-7)

$$L'' = \frac{2\rho C_L^2 \left(\frac{\alpha\lambda}{2\pi}\right)}{\left[1 + \left(\frac{\alpha\lambda}{2\pi}\right)^2\right]^2}$$

(6-8)

$$L = L' + iL''$$

(6-9)

$$tg\delta = \frac{L''}{L'}$$

(6-10)

where

ρ is material density,

C_L is longitudinal sound velocity,

α is attenuation,

λ is wavelength which can be calculated as a function of frequency.

Then the formula for L' can be simplified to $L' = \rho C_L^2$ because the effect of the attenuation is very small and can be neglected. The extent of the reaction parameter, α_{US}, was calculated as:

$$\alpha_{US} = \frac{L' - L_0'}{L_f' - L_0'} 100$$

(6-11)

where

L_f' is bulk longitudinal modulus for the completely cured adhesive, and

L_0' is the modulus value for the uncured adhesive.

Acoustic characteristics of the epoxy adhesive were investigated as a function of cure degree which was reached by heating the sample up to, and cooling it down from, successively increasing temperatures, as shown graphically in Fig. 6-2. Each heating phase provides a different cure stage, and each cooling phase allows for the investigation of the characteristics of the acoustic properties, as a function of thermal cycling, of the system in this particular cure stage. All acoustic and elastic parameters were calculated according (6-1), (6-5) and (6-7) through (6-10).

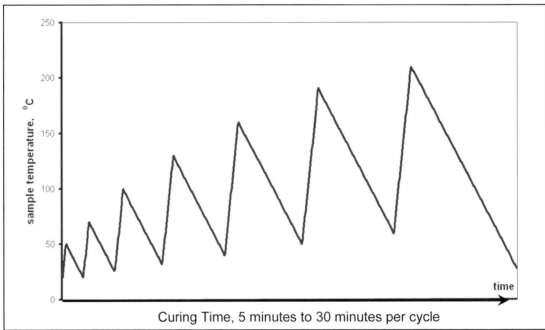

Fig. 6-2. Temperature profile for investigation of temperature dependence of the epoxy on different cure stages.

To acoustically evaluate the effect of curing epoxy adhesive in sheet-metal joints, the adhesive was applied between two metal sheets that were 0.8 mm thick. The thickness of the adhesive layer was maintained between 0.3 mm and 0.5 mm. Specimens were cured from 5 minutes to 30 minutes, at a temperature of 180 C, to reach different degrees of cure. Acoustic measurements were then performed at ambient temperatures, acquiring acoustic velocity data for the adhesive by measuring times of flight between reflected pulses from the upper and lower adhesive-metal interfaces, and using the thickness of the adhesive layer determined by mechanical measurements to then compute the velocity by equation (6-1).

Observation of interface microstructure was made nondestructively with high resolution scanning acoustic microscope (Sonix HS-1000) at the 25 MHz frequency. Pulse-echo mode was used, where the same transducer emits the pulse and receives the echo signal reflections from the interfaces and converts them into electrical signals. This mode is specifically useful for investigation of the material's internal structure. The received signal, acquired by A-scan, is analyzed, and the amplitude of the proper reflection signal was converted to brightness at the corresponding pixel on the acoustic image. In order to obtain a cross-sectional acoustic image along the bond joint, the transducer was moved mechanically across the specimen while sending and receiving the signal at each equidistant position. A C-scan is formed by collecting signal information only inside the user-defined time-of-flight window and moving acoustical lens in two dimensions. Thus, the C-scan represents a horizontal cross-sectional image of the specimen at an operator-specified depth.

By changing the position of the gate, it is possible to obtain images of the internal structures located at a different depths. The acoustic lens was focused at the metal/adhesive interface where the reflections from the boundaries between materials with different acoustic impedances return to build acoustic images. The physics underlying this reflection phenomenon was presented in chapter 3 on theory. The reflection coefficient, R, depends on the difference between the acoustic impedances of the two adjoining materials at the interface according to equation (6-4). Since the properties of the steel (material) remain constant during the process of adhesive cure, the variations in reflection coefficient values can only be attributed to the changes in the acoustic impedances of the adhesive. These changes are caused by modification in either density or sound velocity. Actually, both parameters undergo changes during cure reactions and work together to produce changes in the acoustic images that indicate changes developing in the adhesive during cure. It was shown experimentally that density increases by 2-4% during cure, and all others changes are due to sound velocity modifications.

6.5 Results and discussion

6.5.1 Temperature dependence of the acoustic properties of completely cured adhesive.

We have investigated temperature dependence of the acoustic properties of the completely cured adhesive. In Fig. 6-3, acoustic velocity and attenuation are plotted against the specimen temperature. At ambient temperature, the adhesive's longitudinal velocity is 2500 m/s. Sound velocity decreases monotonically as temperature increase. Analysis of the velocity curve shows a slight change of the slope at 80 C. A similar slope change was also observed in the specific volume vs. temperature curve. This change usually represents the glass transition region where the polymer transforms into its glassy state. Nguyen et al. [189] demonstrated that transition observed at ultrasonic frequencies involves the same relaxation process as the glass transition which is usually characterized by low-frequency techniques like the dynamic mechanical method. The attenuation curve exhibits a peak at 140 C.

6.5.2 Adhesive cure monitoring

Figure 6-4 shows longitudinal velocity and ultrasonic attenuation variations during the adhesive cure at temperatures 120 C and 180 C. At 120 C, the velocity curve has a short lag-period which corresponds to the increase of the molecular weight of the epoxy oligomers. Then velocity increases rapidly at the gel point and, following a reduction in the rate of increase, it then reaching a plateau by 170 minutes. The lag period is not observed at 180 C, instead, the velocity curve increases rapidly in sigmoidal manner, with slope increasing with cure temperature until it reaches a plateau at 30 minutes.

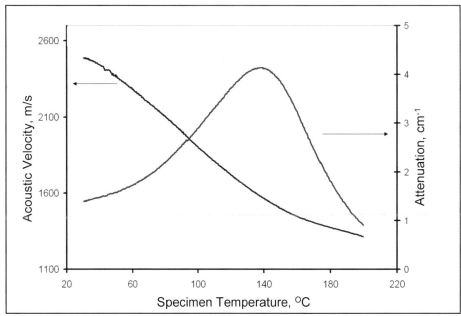

Fig. 6-3. Temperature dependence of longitudinal acoustic velocity and attenuation of the completely cured adhesive.

Attenuation shows its peak in the beginning of the process and decreases afterwards. The peak in attenuation corresponds to the maximum slope in the sound velocity curves for both temperatures. Accordingly, Lionetto [191] and Alig et al. [192], show the peak in attenuation occurs in either gelation or vitrification. Because the temperature of curing is higher then the Tg of the cured adhesive, and much higher that of un-cured adhesive, the peak in attenuation corresponds to the gelation process where the maximum of the epoxy oligomers chains are formed, following the formation of the three-dimensional epoxy network, which leads to decreasing attenuation.

The attenuation maximum, and accordingly gelation time, comes earlier at the higher temperature and the peak itself is more narrow. Earlier arrival of the attenuation peak and its steeper decreasing indicate earlier and faster gelation of the adhesive. A small decrease in attenuation, after reaching the maximum value at 120 C, is related to the relaxation peak for cured adhesive at this temperature. Similar temperature-dependant properties of the epoxy adhesive during cure reaction were observed by Fremantle and Challis [193]. Note that a higher final attenuation value for 120 C than for 180 C results from temperature-dependant properties for cured adhesive (Fig. 6-3) which shows an attenuation peak at the region of 120-150 C.

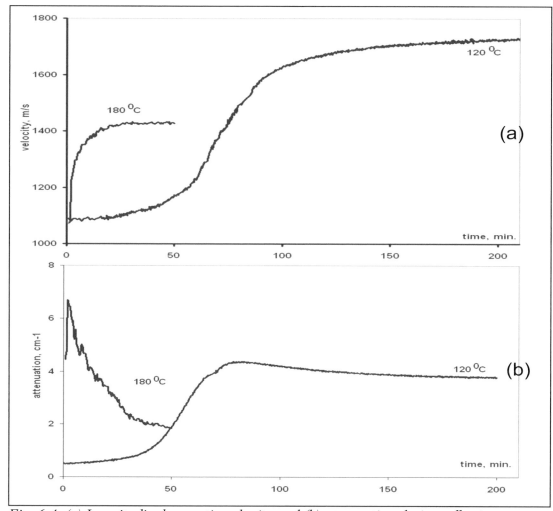

Fig. 6-4. (a) Longitudinal acoustic velocity and (b) attenuation during adhesive cure at temperatures of 120 C and 180 C.

Values of acoustic velocity and attenuation were used for bulk longitudinal modulus calculation. Figure 6-5 shows development of the storage (L') and loss (L") longitudinal modulii and loss factor tan δ (Fig. 6-6) during cure. Maxima in L' and tan δ occur slightly later than the peak in sound attenuation. Pindinelli et al. [188] has shown that the most significant changes in longitudinal modulii occur at gelation. According to Flory's theory of rubber elasticity [194] and Macosko and Miller [195, 196], modulii L correlates rather with crosslinking degree than with the extent of the reaction. The former parameter represents the number of joints in the network which reflects the cohesive properties of adhesive.

The acoustic measurement of progress and extent of the reaction is shown in Fig. 6-7. The final sound velocity and attenuation values correspond to the values of completely cured material at both temperatures, indicating that reaction extent has reached 100% completion. It is worth reiterating for emphasis that α_{US} is based on the longitudinal modulii that represents the degree of crosslinking, rather than the degree of the reaction completion, or functional group consumption.

151

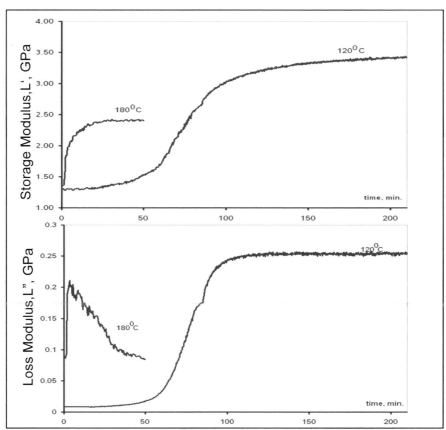

Fig. 6-5. (a) Changes in storage modulus, L', and (b) loss modulus, L", during cure reaction at different cure temperatures.

The loss tangent tan δ value was measured as a function of different cure degrees which were reached by heating the sample up to, and cooling down from, successively increasing temperatures and this data is represented in Fig. 6-6. Initial uncured adhesive has a tangent δ maximum at the 32 C temperature. As the reaction temperature increases, this parameter increases gradually until the cure temperature reaches approx. 130 C. Then, rapid growth of tg δ was observed. After temperature of cure reaches 170 C, the peak temperature becomes a constant 142 C. Maximal increase of loss tangent peak temperature could relate to the gelation point of cure and epoxy network formation. Therefore, transition reflects structural changes during cure of the adhesive and internally linked to the cohesive properties of the adhesive joint.

Acoustic parameters of the epoxy were measured as a function of cure stage which was reached by heating the sample up to and cooling down from successively increasing temperature. Results are presented in Fig. 6-8. Data for the changes in temperature dependence of loss factor tan δ are represented in Fig. 6-9 (a). As the maximum of cure temperature increases, the peaks in attenuation and loss factor shift to the higher temperatures. One can see from both graphs that adhesive's properties change in a stepped manner. Fig 6-9 (b) shows maximum of the tan δ plotted against the cure temperature T_{cure} max. Maximal shift in properties occurs when maximal temperature of cure T_{cure} max reaches 140 C. Note that he peak in temperature dependence of loss factor tan δ, which characterizes glass transition in adhesive, correlates closely with sound velocity in the epoxy which is represented in Fig. 6-10.

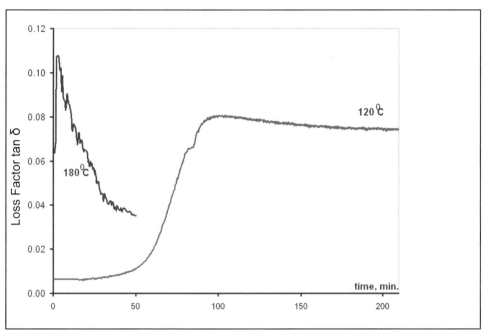

Fig. 6-6. Changes in loss factor tan δ during adhesive cure.

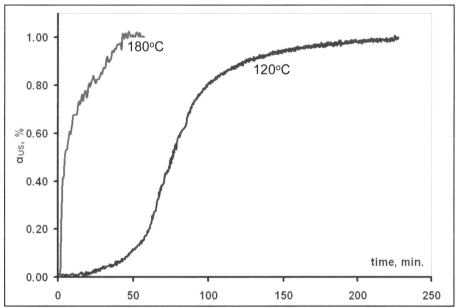

Fig. 6-7. Acoustic measurement of the progress and extent of cure reaction.

Thus, transition of the epoxy from viscous liquid (gel) into solid is accompanied with significant changes in elastic modulii of the adhesive and, consequently, in its acoustic properties. Measurement of the acoustic parameters of the adhesive can be used for in-line non-destructive cure monitoring and quantitative evaluation of the crosslinking degree and cohesive properties of the adhesive.

The acoustic velocity in the adhesive was evaluated at different times in the adhesive bond joints that were exposed to the cure temperature of 180 C, thus the time-dependant velocity data was acquired as the adhesive reached different degrees of cure. Lap shear tests were then performed on these bond joints and the resulting data shown in Fig. 6-11. Acoustic velocity increases more gradually in the beginning, when compared with the curve in Fig. 6-4(a), which can be explained by a slower heating rate of the adhesive in the joint. The maximum value of the velocity is reached by 22 minutes, which agrees with the previous data. Velocity changes reflect variations in the degree of cross-linking of the adhesive during the cure reaction. Joint strength develops during cure in a manner that increases drastically from 0 up to 5 MPa during only 4 minutes of additional curing, from 16 to 12 minutes, and reaching a maximum value of 6 MPa by 20 minutes of total curing time. Adhesion strength develops after gelation during the final stages of cross-linking.

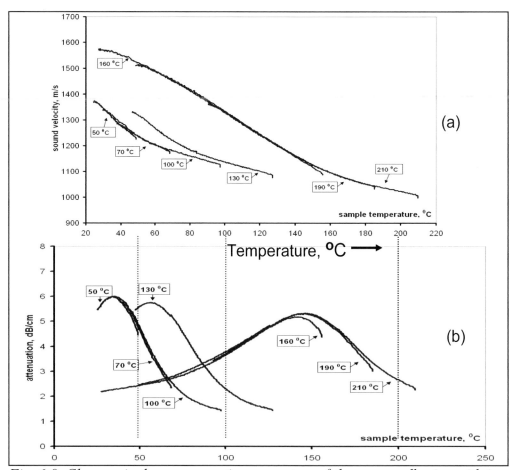

Fig. 6-8. Changes in thermo-acoustic parameters of the epoxy adhesive with increasing cure temperature: (a) acoustic velocity, (b) acoustic attenuation

It is also valuable to note the bimodal distribution of the strength data plotted for the right curve in Fig. 6-11. The bimodal distribution of these strength data was also shown graphically in the histogram plotted in Fig. 1-4(c), where this bimodal characteristic of the distribution of adhesive bond strengths was presented in chapter 1 during the discussion of kissing unbonds and kissing bonds. Now it can be seen that a population of uncured, partly cured and fully cured adhesive bond joints can also exhibit this type of population distribution. It can also be seen from the data analyzed in this chapter and presented in Fig. 6-11, that weak bonds caused by poor curing transmit ultrasound and therefore are not detectable by a simple pulse-echo technique. Such bonds will be ultrasonically identified as bonded, although they are weak bonds, because the poorly cured adhesive wets the surface of the adherend and transmits the ultrasound through it according to equation (6-3). Moreover, experience with the nondestructive and mechanical evaluation of adhesively bonded assemblies from mass production lots has shown

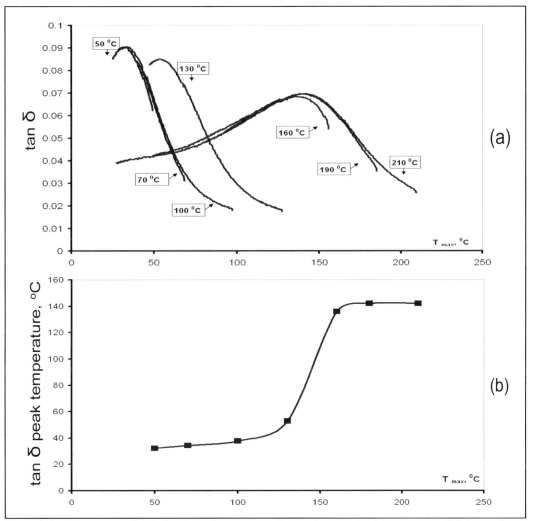

Fig. 6-9. Changes of (a) the loss factor tan δ maximum during cure and (b) tan δ vs peak temperature during cure.

similar bimodal distributions of bond-joint strengths, along with corollary observations of bimodal bond performance and durability. When bond joints survive the first few weeks, they generally survive for the service lifetime of the assembly substrate.

6.5.3 Acoustic imaging of the metal-adhesive interface.

Changes in the material's elastic properties can be visualized on the acoustic images of metal/adhesive interface. Fig. 6-12 shows the development of the microstructure in the epoxy adhesive-metal interface during cure at 180 C. The uncured sample had undeveloped microstructure with a small difference in signal intensities. After having been exposed to the cure conditions, adhesive images looks less homogeneous and reveal granular structure which become more and more developed as cure proceed. Higher velocity in the cured adhesive is

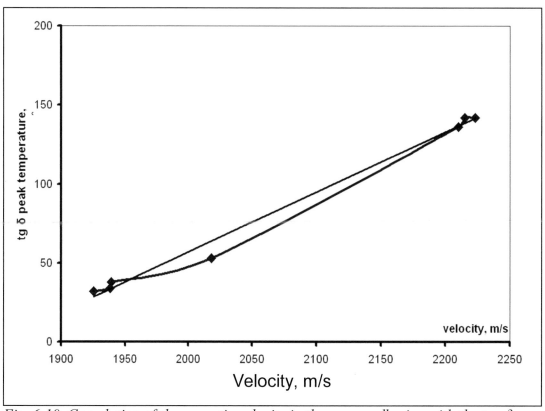

Fig. 6-10. Correlation of the acoustic velocity in the epoxy adhesive with the tan δ maximum temperature.

related to a stronger contact-three-dimensional network adhesive that has increased rigidity, or bulk modulus. Adhesive which was exposed to higher than the recommended cure temperature, in this case 200 C, becomes more brittle and the adhesion strength decreases. Acoustic images of the over-cured adhesive-substrate interface becomes more fragmentary and starts to reveal regions with higher amplitude of the reflected signal which indicates stiffer material and usually corresponds to some physical and chemical changes in the adhesive and/or appearance of micro-defects at the interface. Micro-structural characteristics of the adhesive can give important insight on the process of adhesive cure inside the joint.

6.6 Conclusions

Acoustic techniques that include acoustic microscopy were applied to monitor cure of the adhesive during reaction and to evaluate cohesive properties of the material in the adhesive joint after cure. Acoustic parameters of the thermoset epoxy adhesive were monitored at different temperatures, and results have shown that the acoustic velocity of the adhesive correlates with the cohesive properties and joint strength. This method can be used to monitor development of the adhesive strength during the cure process and to nondestructively detect insufficiently cured

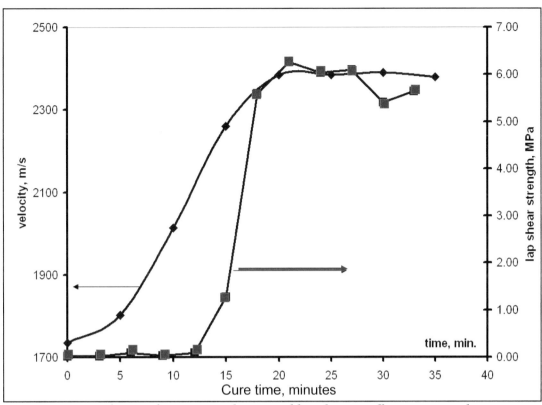

Fig. 6-11 Dependence of acoustic velocity and bond-joint adhesive strength on curing time at the recommended cure temperature of 180 C.

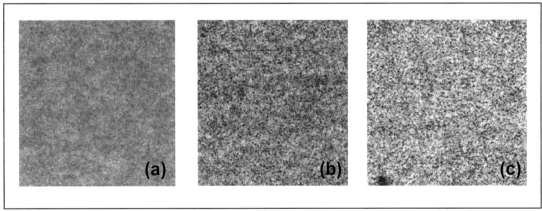

Fig. 6-12. Acoustic micrograph acquired by a 250 MHz scan over a 10 x 10 mm area, showing microstructure of the metal-adhesive interface for (a) uncured adhesive, (b) properly cured adhesive and (c) over-cured adhesive.

joints immediately after the cure stage in manufacturing. Furthermore, these results provide valuable insight into alterations in acoustic propagation when the application of ultrasonic NDE techniques to adhesive bonds in production environments are undertaken before the bonds are cured to the final state. Such situations may not require monitoring the state of cure of the adhesive in the joint, but the effectiveness of the application of the aforementioned technique is greatly enhanced by an understanding of the effect of the uncured adhesive on the propagation and reflection of the ultrasonic signal.

157

7. Development of a 25 KHz Lamb-Wave Bond NDE Technique

The use of fiber-reinforced plastic (FRP) is increasing in the automotive industry in order to provide an acceptable cost-effective approach to accomplish weight reduction for improved fuel efficiency. This increased use of broader material options and FRP assemblies in aircraft, car and truck bodies has resulted in more reliance on adhesive bond joints, in order to provide joining technology for a wide choice of materials in a variety of joint combination. Adhesive bonding also expands design and assembly flexibility while providing components that meet requirements enumerated in the introduction.

7.1 Need for evaluating adhesive bond joints in polymer sheet assemblies

In many of these plastic, FRP, aluminum and multi-material vehicle components, the performance, service life, and appearance of the assemblies are, to a significant degree, dependent on the integrity of their adhesive bonds. Therefore, a method of inspecting bonds for defects is necessary to assure the desired durability and quality of these products. Adhesive voids in FRP assemblies in production vehicles are pictured in Fig. 7-1 as visible evidence of this inspection need. Adhesive bond-joint defects that are not visible in cross-sectional views can be

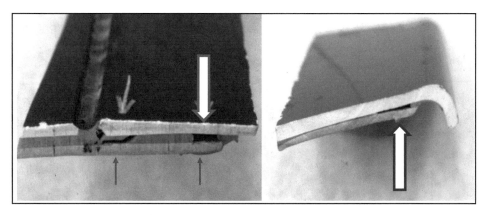

Fig.7-1. *Adhesive voids concealed in FRP bond joints of a vehicle assembly*

illustrated by the unbonds which are identified by arrows in the picture of an opened defective bond joint shown in Fig. 7-2. This photograph shows visible evidence of three of the five possible joint failure modes. The five possible failure modes considered are: (1) unbond failure due to voids and kissing unbonds, (2) adhesion failure due to weak adhesion and kissing bonds, (3) cohesive failure of the adhesive material, (4) delamination of the substrate material where the adhesive is bonded to the adherend surface, but that surface tears away from the substrate below (sometimes called interlaminate failure), and (5) tensile failure of the substrate at or near the bond joint. The three failure modes illustrated are adhesion failure, cohesion failure and delamination of the substrate. The adhesion failures are identified by a clean separation of the adhesive from the adherend surface or substrate. The cohesive failures are identified by the colored adhesive remaining adhered to both substrates. Bond failures by delamination of the substrate are identified by a layer of the adherend substrate material remaining attached to the adhesive layer, as can be seen in the upper right of the portion picture.

An effective nondestructive bond-joint evaluation technique is needed to detect both the lack of adhesive, a cause of adhesive voids, and the lack of adhesion, a cause of kissing unbonds.

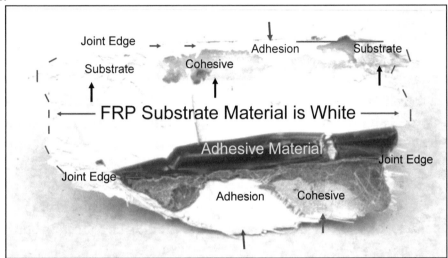

Fig, 7-2. *An opened bond joint showing three failure modes indicated by arrows. (line segments outlined the substrate edges where definition by contrast is lacking.)*

7.2 Requirements for evaluating adhesive bond joints in polymers

An inspection method for use in a high-volume, low-cost production environment must be fast, convenient, cost-effective, and able to effectively assure some minimum acceptable level of bond integrity. It must also be operationally simple and virtually maintenance free. An effective method must provide inspection for adhesion within the bond joint, rather than for the mere presence of adhesive material. The prevailing requirement here was for a method of inspecting for the presence of adhesion, rather than for the mere presence of the adhesive material. There was little emphasis on evaluating the cohesive strength of the adhesive layer, because less than one percent of the failures observed in typical automotive bond joints were cohesive, and in many cases where cohesive failure did occur, the failure load exceeded the minimum load specified for the bond joint. This low fraction of cohesive failures may be due to well-established chemical mixing and thermal curing processes used in these manufacturing operations. Therefore, proven methods of evaluating the cohesive strength of the adhesive material in the bond joint [197] did not provide the solution to the prevailing problem in the application where cohesive failure was an insignificant fraction of the failure modes observed.

There are several other approaches to the NDE of adhesive bonds in polymer composite assemblies, other than the one presented in chapter 5. Each approach considered meets some of the requirements for each specific application. Each offers advantages and has limitations. Some limitations to the ultrasonic pulse-echo method arise from resistance to its implementation in production applications for evaluating the bonds in plastic and polymeric composite components, because of concerns about one key technical issue and two valid manufacturing issues.

The key technical issue was addressed in chapters 1, 3 and 4 during a discussion of the fundamental cause of the effectiveness of the pulse-echo NDE method. That fundamental cause of a reflected echo is marginalized in applications where there is only a small difference between the acoustic impedance of the substrate material and that of the adhesive. This reduces the reflection coefficient at the interface between these two materials, as discussed in chapter 4, where the resulting marginal echo amplitude was modeled. Because this is often the case when polymer composites or plastic components are bonded with a polymeric adhesive, the search for a method was consequently directed toward the low-frequency Lamb-wave approach, examined in chapters 3 and 4, as a reliable, cost-effective ultrasonic method of evaluating adhesive bond joints.

The two valid manufacturing issues are: (1) the use of a liquid couplant on the pre-painted surface of assembled parts. Liquid couplant required for typical high-frequency ultrasonic inspection is often resisted by manufacturing operators and considered undesirable on polymeric surfaces, because of the possible adverse effect on subsequent processes needed to produce the product. Moreover, the use of a liquid couplant on the painted surface of finished assemblies is tolerated, but not welcomed. (2) Inspection speed required for synchronization with production. Therefore, having considered these two manufacturing concerns and the aforementioned fundamental technical issue, a method of NDE was sought that alleviated each concern, while providing effective results at reduced, but acceptable, resolution.

7.3 Lamb wave propagation is an effective method for polymer joints

Among the NDE approaches considered, a 25 kHz bond test method, based on the propagation of Lamb waves along the bond joint, was determined to hold the highest potential for meeting the requirements addressed above. Such a method would need no liquid couplant, is operationally simple and can be performed with portable instruments. Because of its operational simplicity and speed, the method would have a high potential for cost-effectiveness. Its flaw-finding effectiveness is dependent on two key parameters developed and reported herein. They provide the basis for a practical semi-quantitative NDE method. They are: (1) local bond integrity (LBI) measurements, which are based on the use of a statistically selected reference specimen for "calibrating" the bond testing instrumentation, and (2) bond merit factor (BMF) values which are estimates of regional bond integrity computed from LBI data. The effectiveness of these two parameters as indicators of bond strength in plastic FRP lap joints was evaluated in this chapter.

As discussed in chapter 3 and modeled in chapter 4, Lamb waves have long been used in the effective acoustic interrogation of laminated media, such as adhesive bond joints. This chapter will present the methodology and data specific to the technique developed.

7.3.1 Principles supporting the use of Lamb waves for evaluating adhesive joints

The physics of Lamb wave propagation was presented in chapter 3 on theory, where Lamb waves were described as acoustic perturbations in elastic media in which the thickness of the plate or sheet medium carrying the waves is of the same order of magnitude as the wavelength. These waves are named in honor of Horace Lamb, because of the fundamental analytical contributions that he made to the subject. Investigations with Lamb and leaky Lamb waves have been carried out since their discovery in areas ranging from seismology to nondestructive testing, acoustic microscopy and acoustic sensors. In an elastic plate or sheet that is sufficiently thin to allow penetration of the propagated energy to the opposite side, say on the order of one wavelength or so, Rayleigh waves degenerate to Lamb waves which can be propagated in either symmetrical or asymmetrical modes. The velocity of propagation is dependent on the frequency, material thickness, density and modulus, or flexural rigidity. Equation (3-39) from chapter 3 is repeated here to reiterate the quantitative relationship among these variables.

$$c = (D_p/\rho h)^{1/4} \omega^{1/2} \qquad (3\text{-}39)$$

where

c   is the wave speed of the fundamental flexural mode,

D_p is the flexural rigidity of the plate,

ρ is the density if the material and

h   is the thickness if the plate.

D_p , the flexural rigidity of the plate, is related to fundamental material properties by

$$D_p = 8\mu(\lambda + \mu)h^3/3(\lambda + 2\mu), \qquad (3\text{-}40)$$

Because this chapter focuses on the use of Lamb waves for nondestructive evaluation of adhesive bonds, the discussion will generally focus on ultrasonic Lamb wave propagation in adhesive bond joints.

In ultrasonic Lamb wave inspection of adhesive bond joints, the waves can be introduced into the bond layers by mechanically coupled transducers, electromagnetic perturbations or excited by laser impingement upon the surface of the joint. In any case, the acoustic, often ultrasonic, wave propagation is guided between two parallel surfaces of the test object, which in this case is the layered elastic media comprising the bond joint. Because acoustic perturbations are fundamentally more responsive to mechanical properties than thermal or x-ray interrogation of the bond joint, low-frequency ultrasonic inspection has been recognized widely as the desired inspection approach for mass-production manufacturing applications. Other approaches can and may be used to compliment and/or confirm the result, but the acoustic approach has the highest potential for yielding reliable results. The low-frequency is recommended to avoid applying a liquid couplant to the surface of the bond joint.

It can be seen from the discussion in chapter 3 on theory and chapter 4 on modeling, that the amplitude of the asymmetric Lamb wave is increased when it is propagated along an unbonded region of the bond joint. Moreover, the velocity of propagation is decreased. Therefore detecting changes in the velocity and amplitude of propagation are the essential indicators of changes in bond state. Changes in amplitude were detected by monitoring the incoming 25 kHz signal on the receiving transducer. Changes in velocity were detected by monitoring a shift in the phase of the 25 kHz signal on the receiving transducer. This is illustrated in Fig. 4-11 in chapter 4, showing output from the modeling of signals from monitoring a bonded region, and the corresponding output from the modeling of signals from monitoring an unbonded region. The modeling is verified by data presented in this chapter.

7.3.2 Operating principles

The twin-transducer probe used to excite and receive the signal is shown in Fig. 7-3. It contains two 25-kHz transducers. The two transducers, separated by a distance of about 1.8 cm, are fitted with small, conical high-density plastic contact tips for acoustic coupling to the workpiece. One transducer excites the Lamb wave that is transmitted along the bond joint and the other receives the transmitted signal, whose amplitude and velocity are determined by the mechanical condition of the transmitting bond joint. The results reported here will be those obtained using these monitoring techniques with the twin-transducer probe, shown in Fig. 7-3, excited by the 25 kHz voltage square-wave train shown in Fig. 7-4. By utilizing a combined phase and amplitude detection circuit, this monitoring combination was effective for detecting and quantifying certain unbond conditions and marginal adhesion described later.

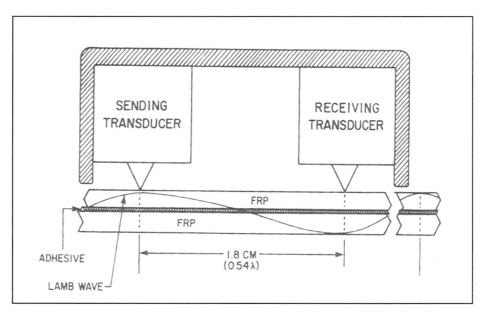

Fig. 7-3. Cross-sectional view of the 25 kHz twin-transducer probe on a bond joint

The asymmetric Lamb wave is initiated in the bond joint by exciting the material under inspection with a series of acoustic pulses from the transmitting transducer. An oscillogram of the electrical pulses that are sent to the transmitting transducer used to excite these 25-kHz wave trains in the test material is shown in Fig. 7-4. The excitation pulses are 4 cycles during 160-ms and have a 20-Hz repetition rate. This pulse travels through and along the material under inspection, exciting 25-kHz vibrations with amplitudes and velocities affected by the local acoustic features of the specimen. The asymmetry of the Lamb wave occurs because the wave is excited by the introduction of vertical perturbations normal the plate that are mode converted to transversely propagated asymmetric waves because the plate thickness of 2 mm to 10 mm is less than the 33 mm wavelength. The signals have also been detected on the opposite side of the plate.

The wave train is propagated along a bond joint five to 10 times thicker than the wavelength, and detected by the receiving transducer. Adhesive bond defects are detected by comparing the wave train of the signals received from the specimen to the wave train of signals received, during previous calibration, from a reference specimen of known bond integrity. The contrast between these received signals can be seen in oscillograms of a bonded region, shown in Fig. 7-5(a), and for an unbonded region in Fig.7-5(b). The significant difference between the amplitudes from the good reference bond and that from an unbonded region can easily be seen by comparing the oscilloscope traces at 125 μs. The 9-μs phase shift manifested by the oscillogram from the unbonded region agrees with the output from modeling shown in Fig. 4-LF3. The three-fold increase in amplitude manifested by the oscillogram from the unbonded region, compared to the amplitude from the bonded region, does not reach the eight-fold increase in amplitude predicted by the model, but several explanations for this deficiency were proposed in chapter 4, the most valid of these being that although the unbonded region may be only half the effective thickness of the bonded region, yet it is not free to oscillate unconstrained as if it were a free plate with half the thickness, especially when bond exist nearby.

This change in phase and amplitude occurs because the pulse transmission is along a layer of unbonded material. The unbonded layer is freer to vibrate at higher amplitude, with less energy dissipation and lower wave velocity than a bonded reference specimen. Lamb waves are also highly dispersive, meaning that different frequencies travel with different phase and group velocities in the medium; therefore it is necessary to acquire evidence to support the hypothesis that the phase shift is due to lower wave velocity rather than to a change in resonant frequency.

That evidence is shown in the acoustic frequency spectra pictured in Fig. 7-6 and Fig. 7-7. These spectra were recorded at a 1-kHz scan width with frequency increasing from left to right. Figure 7-6 shows the frequency spectrum from a bonded specimen, as detected by the receiving transducer. The distribution, over 95 percent of which lies between 24.6 and 25 kHz, is nearly symmetrical about a 24.8-kHz center frequency. Figures 7-7(a) and 7-7(b) show spectra recorded while monitoring unbonded specimens. The upper spectrogram, Fig. 7-7(a), recorded while monitoring a specimen producing a low phase shift, has a bimodal distribution with the higher frequency peak at 24.93 kHz. Its bandwidth, however, is approximately the same as that shown in Fig. 7-6 for a bonded specimen.

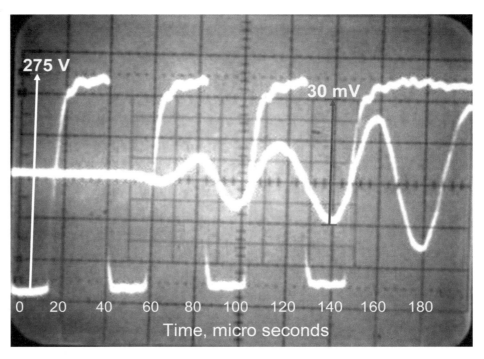

Fig. 7-4. Oscilloscope traces showing 25 kHz square wave pulse for transducer excitation and resulting transmitted wave train received

164

The lower spectrogram, Fig. 7-7(b), recorded while monitoring a specimen producing a much higher phase shift, shows a distribution with bandwidth and symmetry similar to that shown in Fig. 7-6. The absence of higher frequency components implies that the concomitant high phase shift is not caused by an increase in resonant frequency. Conversely, the higher frequency components observed in the bimodal distribution were not assessed to be responsible for the concomitant low phase shift, because a higher phase shift was observed at the unperturbed frequency. Although the correspondence between frequency distribution and phase shift shown in the contrasting spectra of Fig. 7-7 was not consistently reproduced for all specimens monitored, the spectra observed either confirmed the hypothesis, or failed to contradict the hypothesis, that phase shift is caused by a decrease in wave velocity rather than an increase in the resonant frequency of the bond joint.

Velocity measurements, for cases where high and low phase shifts were observed, confirmed that phase shifts were due to a significant velocity decrease. The decreases in velocity from a bonded to an unbonded specimen ranged from about 35 m/s, yielding a 1 μs shift in phase, to 30.6 m/s, yielding a 9 μs shift in phase. The larger phase shift, and concomitant lower velocity, being attributed to wave transmission in a plate of half the thickness, as discussed in chapter 4. The velocity change was determined experimentally by measuring changes in wave arrival times from oscilloscope records like those in Figs. 7-4 and 7-5. The velocity was measured by a similar method. However, another 25-kHz transducer, separate from the two-transducer probe, was used as the receiver. It could be moved along the bond joint to vary the distance, Δx, between the sender and receiver. The wave or phase velocity was determined by measuring the difference in time, Δt, between the arrivals of corresponding phase angles in the incoming wave train, at two positions of the receiving transducer. The velocity was computed, using data from several such measurements, as 833 m/s. The wavelength was measured with the same transducer arrangements, and also computed from velocity and frequency data. Both methods yielded 3.3 cm for the FRP material used in this investigation.

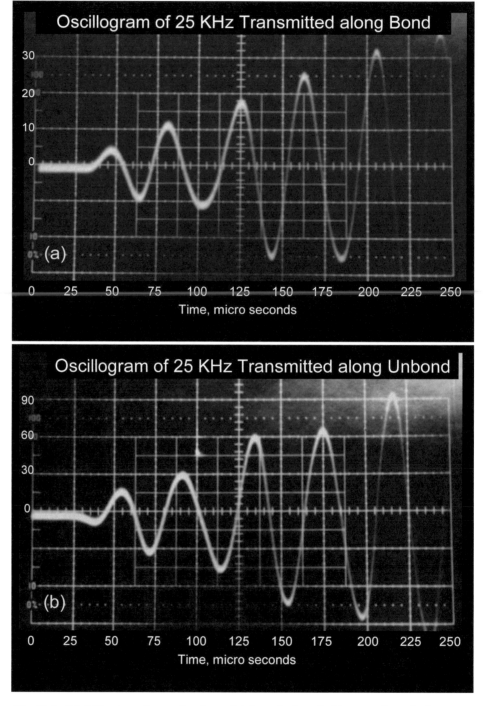

Fig.7-5. 25 kHz signal received on (a) bonded and (b) an unbonded region of a joint. The 9-μs phase shift, due to lower transmission velocity for the unbonded case, can be easily seen as the peak at 125 μs in (a) shifts to 134 μs in (b).

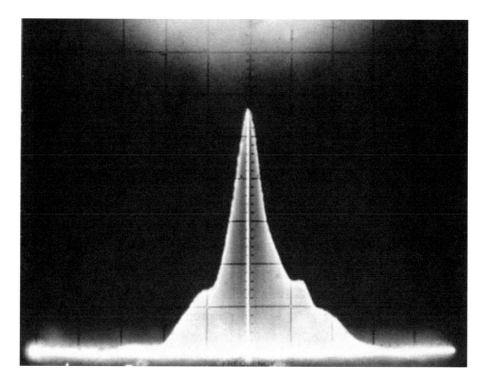

Fig. 7-6. Lamb-wave frequency distribution for a 25 kHz excitation of a bonded region of the FRP joint, registering no phase shift.

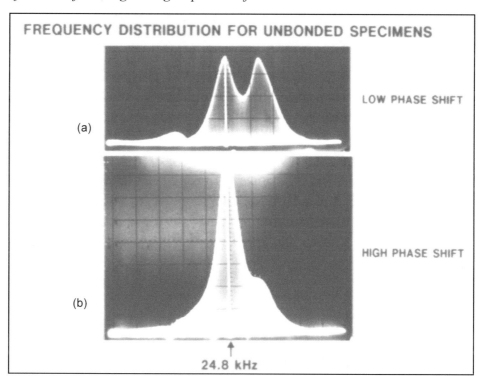

Fig. 7-7. Lamb-wave frequency distribution for 25 kHz excitation of an unbonded FRP joint region that registered (a) a low phase shift and (b) a high phase shift.

7.3.3 Acquisition of quantitative data

In order to make a valid comparison of pulses from bonds of unknown integrity with pulses from a reference specimen of known integrity, the bond tester was first referenced, using the required bonded specimen to calibrate amplitude and phase levels. The two options for detecting unbonded regions are (1) flaw detection by sensing deviations of the phase from its reference value and (2) flaw detection by sensing the combined deviations of the phase and amplitude. This was the option used throughout this study. The alarm activation level (AAL) adjustment, made after the amplitude and phase levels are adjusted, determines the turn-on point of a Schmitt trigger which activated an alarm. Its setting determines the magnitude of increase in amplitude and shift in phase required for flaw indication by alarm activation.

In addition to the alarm mode for which the bond tester was designed, the instrument's use in this study was extended to a more quantitative mode: that of reading the instrument response to the acoustical property of the local bond region. This attempt to quantify bond tester response was motivated by the assumption that these responses may indeed correlate with bond strength, or at least to some degree with bond joint performance.

The responses of interest were those which combined amplitude change and phase shift. The AAL was available to do this. It was adjusted during the stationary monitoring of a local region, until marginal alarm activation occurred. This AAL potentiometer reading was then recorded as the NDI datum for that bond locale. This was an awkward reading to take because it required adjusting the AAL potentiometer before each reading, until marginal alarm activation, and operator judgment to decide when marginal alarm activation occurred. Marginal alarm activation here was judged to occur when the alarm was intermittently on as much as off. Because of the time-inefficient awkwardness and possible operator error involved in taking AAL readings, another method was devised for reading an equivalent indicator. The bond tester alarm circuit, a portion of which is shown in Fig. 7-8, was altered so that the potential difference between the base of transducer Q19 and ground could be measured with digital voltmeter M4. The AAL potentiometer, R87, was set to read zero (at 200 ohm maximum resistance) during the NDI monitoring, and the voltage readings were recorded as LBI readings for each bond locale.

The functional relationship between LBI readings and AAL values was explored to confirm the validity of this approach to improving bond data acquisition. The LBI versus AAL data for each point were obtained while monitoring a given bond location, under constant probe pressure applied by the 9.8-Newton force exerted by a 1-kg mass in a well-controlled laboratory environment. The correlation coefficient of the data was 0.9999, with an excellent closeness of fit. This indicates both the precise relationship between LBI and AAL values, and the high precision obtainable from the instrument when small variations in probe position and pressure are not contributing factors. This empirical function is one member of a family of such curves, one of which exists for each referenced level of instrument operating sensitivity.

The LBI voltage readings are a linear function of the AAL network resistance, because an obvious E = IR type linear relationship exists, where E is the LBI voltage, I is the current through

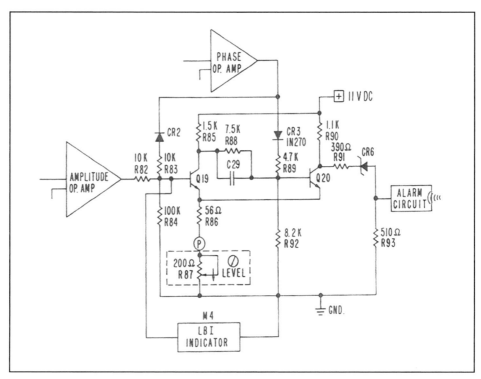

Fig. 7-8. Circuit schematic for acquiring local bond integrity readings

the AAL network (R84, R86, and R87) shown in the circuit diagram, and R is computed by

$$R = \cfrac{1}{\cfrac{1}{R84} + \cfrac{1}{R86 + R87}} \qquad (7\text{-}1)$$

A plot of this function is not shown because the obvious trivial nature of the relationship, and it is similar to a plot of LBI voltage versus the AAL potentiometer resistance, because R84 is much greater than the sum of R86 and R87. Hence the correspondence between LBI and AAL was reliably established.

7.4 Bond Joint Specimens

Specimens were prepared from FRP plaques by bonding two 10 by 31-cm plaques with Goodyear Pliogrip 6000, a two-part urethane adhesive system. The plaques were bonded to form a 3.5-cm-wide lap joint, 31 cm long. The adhesive thickness was maintained at 0.8 mm. The bond joints contained a combination of adhesive voids, unbonds, and other intentional substandard regions. The bonded plaques were marked to uniquely identify each 2.5-cm segment along the bond for cutting and destructive testing subsequent to NDE.

The plaques for 50 of these specimens were made by compression molding a polyester sheet molding compound (SMC) containing 25 to 30 percent glass-fiber reinforcement. Plaques for 22 of the specimens were made by compression molding a high-glass content polyester sheet molding compound (HSMC) containing about 40 percent glass-fiber reinforcement. The SMC specimens were prepared to represent non-loadbearing bonded assemblies, and the HSMC specimens were prepared to represent bonded structural assemblies.

7.4.1 Bond reference specimens and procedure

In order for the evaluation results to have reliable consistency, bond reference specimen must be generated. The proper preparation and selection of a bond reference specimen is a prerequisite to obtaining meaningful and consistent results in this approach to the NDE of bond joints. A reference specimen is required for use as a calibration standard because the bond tester essentially operates as a difference-detecting device. It compares the acoustic characteristics of unknown parts with those of a reference of known acceptable integrity. It is, therefore, essential that the reference specimen resemble the parts to be inspected in material composition and bond joint geometry. It should also possess a level of bond integrity equal to or only slightly better than that required in the final product.

A reference specimen was selected, for each of the two material types, from sample sets which were similar to the corresponding test specimens in geometry and composition. Each sample set of about 25 bonded specimens was prepared under adhesive bonding conditions that closely simulated accepted and expected production practice. These specimens contained no intentional adhesive flaws. Or substrate voids. Each reference specimen is selected from the sample set by the following procedure:

1. Use one arbitrary bonded specimen as an interim reference to initially adjust the bond tester according to the level of sensitivity desired [198]. During this referencing, and all subsequent measurements in the procedure, each specimen should be supported by a material of low acoustic impedance, such as soft plastic foam sheet about 2 cm thick, to reduce acoustic interferences from the supporting structure.

2. Monitor several bond joint locations on each of the specimens by placing the probe down firmly on a location along the bond line. A 1-kg mass is recommended for use as a probe hold-down weight, to assure the application of evenly distributed, reproducible and operator-independent firm pressure on the probe. (About four

locations per specimen are suggested.) Avoid locations where sharp changes in the readings are observed for small changes in probe position. Record phase, amplitude, and AAL values for each monitored location. (LBI readings may be recorded instead of AAL.)

3. Permanently identify (mark) each location and indicate the transducer probe orientation for future reference and duplication.

4. Rank each monitored location according to its AAL value. Phase and amplitude values are recorded to verify that the arbitrary interim reference specimen was indeed bonded. AAL values are the combined results of phase and amplitude values and are used here to indicate bond integrity.

5. Select those bond locations ranking at or near the 40th percentile as reference candidates. These reference candidates should have bond integrities, as indicated by their AAL values, higher than 35 percent but no higher than 45 percent of the AAL values at bond locations monitored in the sample set. Each specimen containing one or more of these locations is a reference specimen candidate.

6. Reference the bond tester on the reference specimen candidate nearest the 40th percentile according to the described procedure.   Then inspect each specimen ranked below the 40th percentile to find the highest ranking specimen causing bond-tester alarm activation.

7. Destructively test the highest ranked specimen causing alarm activation, and its two nearest neighbors according to a prescribed mechanical test or durability test procedure, to determine whether they met engineering bond specifications.

8. If the bond strength meets minimum requirements for the intended adhesive system application, the candidate specimens from the 40th percentile are confirmed as valid initial reference specimens for the subject adhesive/adherend assembly application.

7.4.2 Specimen Support Box and Probe hold-down force

Because the Lamb wave flexural mode requires a freely unrestrained plate for propagation, a specimen support box was provided to support the bonded plaques during NDE monitoring. The box was made of corrugated paper with half-inch thick soft plastic foam sheet covering the edges supporting the plaques. The dimensions of the box were such that only the extreme and edges of the plaques were in contact with the foam padding. The box supported the specimens in a manner which disallowed significant acoustic interference by a supporting structure.

A 1-kg mass, weighing 9.8 Newtons, was used as a probe-pressure weight to hold the probe down on the specimen during monitoring. This removed one component of operator variance and assured a constant probe pressure during the monitoring of each specimen.

7.5 Nondestructive Evaluation of Specimens

7.5.1 Referencing

Three adhesively bonded FRP lap joint specimens were used to perform the referencing adjustments. Two of these were reference specimens of known acceptable bond integrity. They were selected from a sample set of bonded specimens according to the procedure described in the preceding Bond Test Specimens section. The third contained an unbonded area about 1-in.^{2} within the lap joint. One of the two reference specimens of sound integrity was kept with the instrument for regular use as a control. The other was retained as a "primary standard" and kept in a protected place for daily comparison to the first. The third specimen was used to confirm unbond detection after completing the referencing adjustments.

These specimens were used to reference the bond tester according to instructions in the instrument operator's manual. To control the operating sensitivity of the instrument, referencing was repeated whenever the bond tester had experienced power interruption, mechanical or thermal shock, been left unattended, or in operation for more than one hour.

Possible bond deterioration of the reference specimen during use is monitored by daily comparison with a second reference specimen, such as the "primary standard." The phase, amplitude, and AAL or LBI readings are recorded and compared daily, or before each use, using the same specimens for each daily referencing. When a significant change in the relative bond integrity of the two references is detected, the reference specimen with waning bond integrity (decreasing AAL or increasing LBI) is discarded and a replacement selected.

Similar reference procedures are used in other instrumental methods of materials characterization [199] where the accuracy or long-term precision (drift) of quantitative measurements must be carefully controlled.

7.5.2 Evaluating bond joints for local bond integrity

The specimens inspected for bond strength were marked for identification and were then suspended across the padded edges of the specimen support box, and the central bond region of each 2.5-cm segment interrogated with the bond-tester transducer probe aligned as shown in Fig.7-9. The probe hold-down weight was placed on top of the probe during monitoring. LBI readings were recorded for each bond segment. Eleven readings were taken along the bond line of each plaque.

7.5.3 Evaluating bond joints for unbonded regions

Bond joints are inspected by moving the transducer probe along the joint, aligned with tips parallel to, and about 1.3 cm (0.5 in.) from, the bond edge. Firm but gentle contact pressure was applied while the probe was moved along slowly, at less than about 15 cm/s

(6 in./s). When an unbonded area was detected by the activation of the alarm, the probe was moved laterally across the bond joint, first to one side and then to the other side of the unbond, to search for nearby bonding. If bonding was located within the area, such that the width of the unbonded region was estimated to be no more than two thirds of the total width of the bond joint, a partial unbond was indicated and identified as such in the data record.

A complete unbond was indicated when the alarm continued over a width greater than two thirds of the total bond joint width, as the probe was moved laterally from one side of the bond joint to the other. Both a partial unbond and a complete unbond are illustrated in Fig. 7-9. The circles in the schematic, representing the transducer contact tips, indicate the alignment of the transducer probe parallel with the joint edge during the entire inspection operation.

The data resulting from the inspection were recorded by each operator in a record for the specimen in such a way as to identify the approximate location and extent of the unbond (its approximate center, length, and whether it is partial or complete). In a production environment, unbond data from each specified bond region of the assembly could be recorded on a schematic of the assembly or tabulated under headings which identify each specified region with a mnemonic set of alpha-numeric characters. The tabular scheme is more desirable in cases where computerized data storage and reduction is expected to be employed. The results will be given in the Results and Discussion section.

In a practical production application, unbond and bond data will be used to accept or reject a bonded assembly or region thereof. It is impractical to assume that any minimum substandard bonding in an assembly would warrant rejection of the part, so a formula was

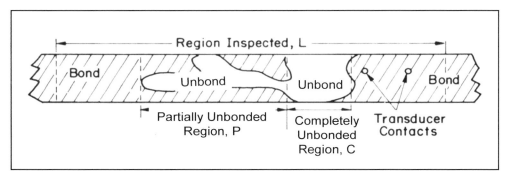

Fig. 7-9. Plan view of lap joint showing a schematic of partial and complete unbonds.

devised to quantify the bond integrity of a bonded region of approximately 25 to 40 cm long. Then the a quantitative estimate of bond quality can be stated for each such region, in terms of local bond inspection data. Unbond dimensions are defined in terms of the operating sensitivity level of the bond tester. These unbond data, which may include areas of weak or substandard bonding, are then used to compute a BMF for the region by

$$BMF = \frac{L - C - P/2}{L} \tag{7-2}$$

where

L = total length of bond joint inspected,

C = total length of all complete unbonds within the region, and

P = total length of all partial unbonds within the region.

Partial and complete unbonds are defined and the inspection process for detecting them was described earlier in this section under Evaluating bond joints for BMF. Figure 7-9 shows an example of partial and complete unbonds.

7.6 Mechanical testing of specimens for correlations with NDE results

The specimens were prepared for mechanical testing by cutting the plaques, described in the Bond Test Specimens section, into 2.5-cm segments with a diamond-tooth saw. Each 2.5-cm-wide segment shown was cut to an overall length of 18 cm. The resulting specimens were then subjected to the following tests to determine bond joint strength.

7.6.1 Shear strength by tension loading

Test methods used were based on procedures described in the American Society for Testing and Materials (ASTM) Recommended Practice for Determining the Strength of Adhesively Bonded Rigid Plastic Lap-Shear Joints in Shear by Tension Loading [6] and Test for Strength Properties of Adhesives in Shear by Tension Loading of Laminated Assemblies [200]. However, the previously described bonded specimens were not identical in geometry to those described in the ASTM standards. The specimens were mounted in an Instron testing machine, model 1115, with a 20,000-lb load frame. Each specimen was loaded to failure at a crosshead speed of 1/27 mm/min, and the load at failure recorded. The failure mode was also observed and recorded.

A test procedure based on ASTM recommended practice D-3163-73 (1979) was used to simulate the tension shear-load conditions experienced by the lap joints while in service. This test also conformed to mechanical tests required by engineering specifications, with which the NDI method was being developed to correlate. It did not measure the pure shear strength of the bond, because immediately upon loading, the lap joint specimen deformed to the configuration shown in Fig. 7-10. This placed the joint under a combination of shear and peeling forces, as well as a flexural moment enhancing tearing of the FRP adherend. An attempt was made to measure a more accurate bond shear strength. The test procedure, based on ASTM test D-3165-73 (1979), was used to increase the shear component at the adhesive-adherend interface by loading along the neutral axis, reduce distortion-induced tearing, and reduce the fraction of failures by delamination within the adherend. Specimens for this test procedure were fabricated to

174

have a dual thickness over the entire span of the specimen, except for a notched indent, on alternate sides to define the lap joint test area. The results of such tests were predominantly FRP adherend failures by delamination at loads similar to those obtained previously. Moreover, this type of specimen, designed to load the bond joint along the neutral axis, is not representative of the loading imposed on bond joints in actual vehicle applications; hence it was not recommended for continued use in this and future investigations for the intended real-life applications.

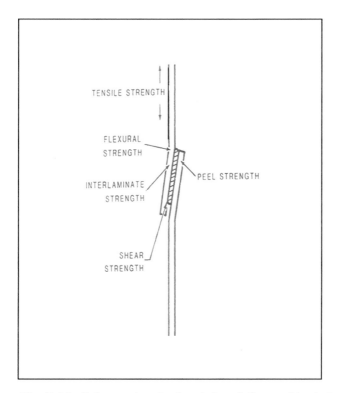

Fig.7-10. Schematic of a lap joint deformed by being under tension load for lap-shear testing. It shows the five components of joint strength tested when such specimens are not loaded along the neutral axis.

7.6.2 Peel strength by flexural testing

A test method based on the procedure described in ASTM Tests for Flexural Properties of Plastics and Electrical Insulating Materials [201] was used. The bonded specimens, with the previously described conventional alp joint geometry, were loaded at the rate of 5 cm/min at the center of the lap joint by a 3-mm-radius loading nose, while the specimens were supported over a 5-cm span. The testing machine used was a Baldwin custom-built prototype with a 60 000-lb load frame.

7.7 Results and Discussion of Bond Test Data

7.7.1 Unbond detection and reporting

This bond NDI mode detects unbonded regions whether caused by lack of adhesive or lack of adhesion. It does not, by the recommended alarm mode, distinguish between unbonds due to adhesive voids and unbonds due to unbonded adhesive. The operator can, however, distinguish between adhesive voids and unbonded adhesive by observing the deflections of the phase and amplitude meters. The phase meter is usually deflected to the left when the unbond is caused by unbonded adhesive. The amplitude meter is usually deflected to the right when unbonds are due to adhesive voids. Furthermore, unbonds detected by the bond tester cannot always be confirmed by x-radiography, because many of them are not adhesive voids. The fact that 83 percent of them in this group of specimens were voids is indicative of the method by which unbonds were intentionally produced in these specimens, and the level of instrument operating sensitivity used to detect them. In actual production, however, voids may be less likely to occur than unbonded adhesive. This has been observed many times.

The probability of detecting regions of unbonded adhesive and adhesive voids increases, as expected, with unbond size. Data to confirm this were obtained from the four-operator inspection of the 50 bonded specimens previously mentioned. Of the 115 unbonds in the sample set, all were wider than 0.6 cm, the minimum unbond detection width. No adhesive voids or unbonds shorter than 1 cm were detected. The probability of detecting adhesive voids varied from about 0.25 at the 1-cm limit of detection, to about 0.5 at 1.5 cm, to 0.95 at 2.0 cm. The probability of detecting unbonded adhesive ranged from about 0.25 at the same 1-cm limit of detection, to about 0.5 at 3.3 cm, to 0.95 at 4.0 cm. The lower detection probability for regions of unbonded adhesive, compared to that

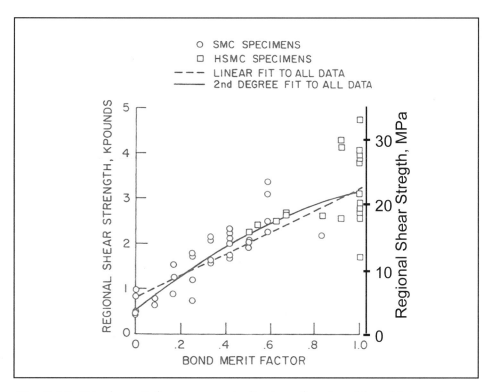

Fig. 7-11. Bond merit factor as an indicator of regional bond integrity

for adhesive voids of the same size, may be due to the unbonded adhesive restricting the oscillation amplitude of the excited adherend layer. This reduces the amplitude of the pulse transmission, which reduces the amplitude component in the phase-plus-amplitude unbond detection circuit.

The minimum unbond detection size was not dependent on instrument operating sensitivity. Increasing the instrument sensitivity did not increase the unbond detection probability for small unbonds, nor did it reduce the minimum size of unbonds detected. The 1.8-cm distance between the transducer tips appears to be the likely limiter of the minimum unbond detection size and resolution. This is implied by the distance between the transducer tips being approximately equal to the 2.0-cm adhesive void size with 0.95 detection probability, and to the 2.2-cm, pooled standard deviation, precision of lineal unbond definition. These distances are approximately 60 percent of the 3.3-cm wavelength of the interrogating acoustic energy. It is therefore unlikely that the long wavelength is responsible for the minimum unbond detection limit, except as it limits the minimum size of the transducers.

7.7.2 Bond performance indicators

The nondestructive determination of bond strength is not intrinsically available by ultrasonic NDE methods; however, bond strength over a finite, well-defined bond-joint region can be implied by this NDE technique when bond states are assumed to be tri-modal. The tendency to have three bond-state modes is confirmed by bond-joint strength data acquired from lap-shear mechanical tests shown in Fig. 7-12, where strength data is plotted against arbitrary local bond integrity (LBI) readings from voltages measured in the circuit shown in Fig. 7-8. Note that these data have two frequency clusters: One data cluster is near 1.4 MPa (200 psi) lap-shear strength, one near 3.5 MPa (500 psi) and the obvious third cluster near 0 MPa, when no bonding occurs. Therefore any substandard adhesion within a 2.5 cm region of bond joint is detected and indicated by the changes in transmitted amplitude and phase detected by the bond tester.

Deriving bond strength information from NDE indicators requires a method of inspection which has been verified by destructive mechanical test results. These results were obtained on the specimens described in section 7.4 on "Bond Test Specimens" by test procedures described in section 7.6 on "Mechanical Testing". The specimens were tested to failure, and the load at failure recorded, along with the observed failure mode. Of the four possible failure modes: adhesion, delamination in the adherend, tensile, and cohesive, only the first three were observed. Examples of each of the three are shown in Fig. 7-2 for a production part, and are shown Fig. 7-13 for test specimens. The most frequent failure was by delamination of the adherend enhanced by distortion-induced tearing, which can be classified as a type of corrupted tensile failure.

AAL Indicator – Early preliminary results indicated the potential of this inspection approach to bond strength determination. Initial positive results were obtained when destructive tests were undertaken to confirm the validity of a reference specimen selected according to the procedure described in the Bond Test Specimens section. The reference specimen was selected from 25 bonded HSMC plaques prepared under production conditions. In this selection process, 104 local bond joint regions were monitored, of which sixty-five 2.5-cm-wide specimens were prepared and tested for shear strength by tension loading to failure. Six of them failed adhesively. Data for these six adhesion failures are plotted in Fig. 7-14. A good linear fit to these data was obtained, with a correlation coefficient of 0.998. But the problem which prohibits practical use of this empirical relationship to determine bond joint strength is the super positioned scattered data from other modes of joint failure. These scattered data were within the same domain, and had a very poor correlation coefficient. The best correlation coefficient was 0.33, obtained for a second-degree polynomial fit. The correlation coefficient of the linear fit was 0.10. These data, from all failure modes, underscore a very real problem encountered when evaluating bonds for strength. Because the inspector cannot nondestructively distinguish joints that will fail adhesively from joints that will fail by all other modes, a method which indicates adhesive strength only is an ineffective indicator of bond joint strength, no matter how good the correlation may be. This reveals the limited extent to which an unaltered commercial bond tester can be an effective indicator of bond joint performance in service.

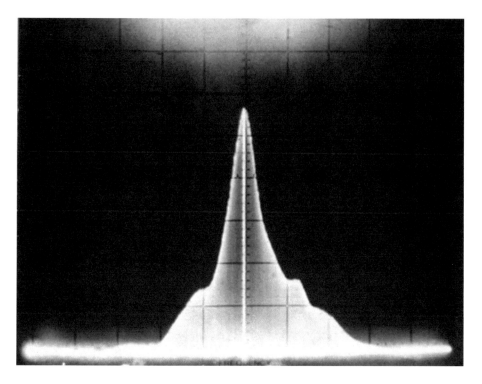

Fig. 7-6. Lamb-wave frequency distribution for a 25 kHz excitation of a bonded region of the FRP joint, registering no phase shift.

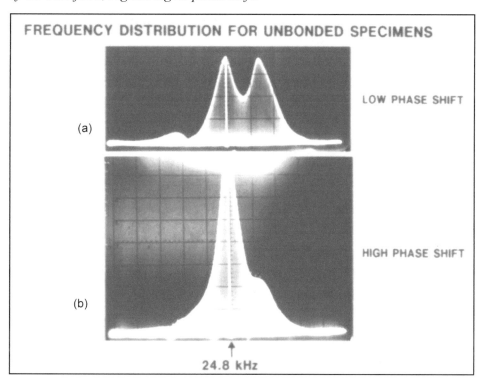

Fig. 7-7. Lamb-wave frequency distribution for 25 kHz excitation of an unbonded FRP joint region that registered (a) a low phase shift and (b) a high phase shift.

When the data in Fig. 7-12 are fit to polynomials of degrees 1 through 4, the relative standard errors of estimate and correlation coefficients are virtually constant over the range of curve fit degrees for adhesion failures, and similarly for data from all failures.

The relative standard error of estimate is less for all fits to the data from all failure modes. The optimum fit was obtained for fourth degree, with a correlation coefficient of 0.91. The reduced correlation appears to result from the reduced slope in the LBI region less than 1.7 V. The reduced slope in this region should not be a serious detriment to the use of the NDI method, because the critical decision region for identifying marginal or substandard bonds in this population is that region where shear strength is less than 2.76 MPa (400 psi). Only about 14 percent of the specimens tested were weaker. Both curves have sufficient slope to indicate usable sensitivity in that critical region. Figure 7-14, where shear strength is plotted against AAL, also show sufficient slope to indicate useful sensitivity in the linear region below 2.76 MPa (400 psi).

These data form the experimental basis for the NDE method developed and reported herein. They do not provide physical insights into the empirical relationships established primarily to provide quantitative interpretation of adhesive bond NDE results with known high correlation coefficients for adhesive bond quality assurance applications.

7.7.3 Correlation of LBI with peel strength is lacking

The correlation of bond peel strength by flexural testing with LBI was poor. The data yielded curve-fit qualities with very low correlation coefficients for several fits to the data from the various failure modes. For 41 tests in which adhesion failure was the predominant mode observed, a linear curve fit gave a correlation coefficient of 0.1. For curve fits with polynomials of degrees 2, 3, and 4, the correlation coefficients were 0.25, 0.63 and 0.63, respectively. These low correlation coefficients, along with high standard errors of estimate ranging 15.5% to 19.3%, indicate no useful relationship exists between the LBI indicator and bond-joint adhesion failure strengths by flexural testing. In fact, the scatter of the data about the best polynomial fit to the 41 data points for adhesive failures was so large that the standard deviation of all the data was approximately equal to the standard error of estimate of the curve fit. This implies a curve fit so poor that about one third of the LBI readings will be outside of the bond-strength range over which the correlation function is defined. The correlation function is therefore virtually useless for practical applications.

The correlation of bond-joint peel strength by flexural testing with LBI was also poor for 28 tests which failed by delamination of the adherend. The highest correlation coefficient obtained for three curve fits to these test data was 0.26, and the lowest standard errors of estimate was 17%. When date from test specimen failure modes which were a combination of the two or three modes were fit to curves with varying degrees, the correlation coefficients were near 0.2 and the high standard errors of estimate ranged from 21% to 24%. These data show that no good correlation or curve fit was found for any of the observed failure modes, or combination thereof, for flexural testing results with LBI values. This agrees with results from the Fokker bond tester reported by Smith and Cagle [202].

These bond-joint mechanical test data underscore the serious defect in a design which loads adhesive bond joints in peel where their strength is weakest. The mead flex-peel strength for all 110 bond-joint specimens tested by this loading mode was 69 MPa (100 psi), far below that required to survive normal performance in a vehicle where lap-shear strengths of 2.76 MPa to 4.14 MPa (400 psi to 600 psi) are the usually specified minimum bond-joint strength requirements.

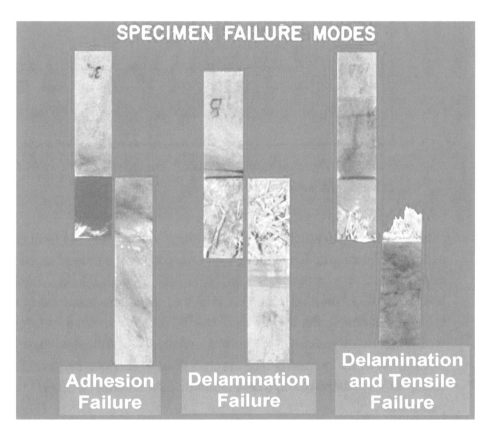

Fig. 7-13. Three frequent bond-joint failure modes in FRP specimens

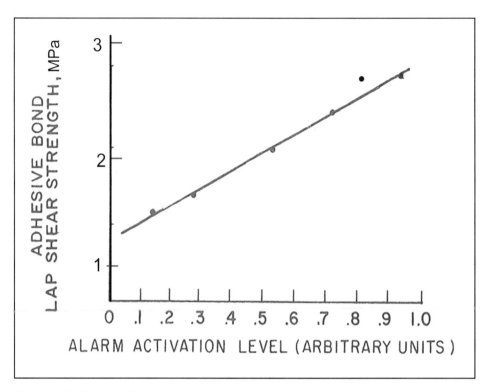

Fig, 7-14. Bond adhesion lap-shear strength vs alarm activation level, for adhesion failures only

BMF Indicator – In routine QC applications, the bond tester is most cost effectively used as a unbond detector. In such use, the AAL potentiometer is adjusted so that alarm activation occurs whenever the local bond joint strength is below a predetermined critical level. This level must be within the region of acceptable sensitivity, indicated by linear or near-linear curve fits in Figs. 7-12 and 7-14. The domain of this region may be different for each specimen population, and must be established experimentally before reliable use of the method can be assured. Past experience indicates that the linear region will generally extend through bond strength data ranking from one to about the 30th percentile.

Once the AAL has been adjusted to the desired level, areas which cause alarm activation during bond joint inspection are identified as regions of substandard bond strength. The extent of these regions is determined as described under Inspection for Regional Bond Integrity in Nondestructive Inspection of Specimens, and a BMF computed by Eq. 2. The degree to which the BMF correlates with regional bond strength is reported herein. These correlations are based on the assumption that the sum of the shear strengths of the 2.5-cm-wide lap joint specimen segments tested approximately equals the shear strength of the plaque, or region thereof, from which they were cut. Under this assumption, data from six adjacent segments were combined to form a bond region datum.

Data from 60 HSMC bond regions were similarly obtained. They are plotted in Fig. 7-11 as squares. They were combined with the previous SMC data for curve fitting. The curve-fit qualities of the combined data are tabulated in the lower half of Table 3. The correlation coefficient of each fit is nearly 0.8. These data, and the resulting curve fits, indicate a useful correlation of regional bond strength with BMF for these adhesive-adherend systems.

The poor precision of the bond-strength data at BMF values near unity occurs mainly because the unbond definition data, from which the BMF values are computed, contain no information on the bond strength of bonded regions; except that it exceeds a minimum critical value. Obviously, there is a wide range of bond strengths that exceed this value, especially among these inspection data which were obtained at a low bond tester sensitivity. If, on the other hand, the operating sensitivity of the bond tester were increased, the sizes and frequency of regions defined as unbonds would increase, and the range of bond strengths existing in the sub-critical or unbonded regions would consequently increase. Lower-range strength data at each low-sensitivity BMF would be shifted to a lower BMF at the high sensitivity, and the poor precision would then shift to lower values of BMF, accumulating at zero, where a higher intercept of the curve-fit would be observed.

The nonzero intercept in Fig. 7-11 is a result of two simple unbond data acquisition features. First, recall that an unbond is called complete even though there may be narrow bonded regions along its edges. Second, poorly bonded regions with bond integrity below the critical level, defined when the bond tester is referenced, will be identified as unbonds. Obviously, these regions always have bond strength equal to, or greater than, zero. The maximum level of bond strength existing in such unbonds is determined by the referenced operating sensitivity of the bond tester.

The correlation of regional bond flex-peel strength with BMF was investigated by inspecting eleven bonded HSMC plaques and cutting them into 110 specimens for destructive testing. This provided 66 regions for BMF computation for correlation with strength. The BMF of these 66 regions ranged from 0.833 to 1.0, insufficient to establish a useful correlation between regional flex-peel strength and BMF. Of these 66 regions, 55 had BMF values of unity. The average strength of these 55 regions, each 15.2 cm long, was 532 pounds, with a coefficient of variation of 24 percent. The coefficient of variation for the strength of all five regions with BMF of 0.833 was 5.3 percent, about a mean of 560 lb. These data tend to show again the poor strength precision at BMF values at or near unity.

7.7.4 Bond-joint geometry

These correlations, established for flat specimens, are assumed to hold for typical bond joints in various vehicular FRP assemblies with a variety of geometrical configurations. The validity of this assumption is based on the small size of the 2.5-cm bond length over which the measurements were made. In most actual assemblies encountered thus far, few bond joints deviate significantly from flatness over such a short distance.

7.7.5 Instrumentation, calibration and operator error and precision

When the results from inspecting the group of 50 bonded specimens for unbonds were summarized, they showed that the effect of inherent operator and instrument variability was virtually insignificant, compared to the size of a typical unbond. The four operators used two bond testers to detect 115 unbonds in this group of specimens. The agreement on location of the lineal center of the unbonds was within 0.7 cm (0.3 in.) pooled standard deviation. The agreement on the length of these unbonds was within 2.2 cm (0.9 in.) pooled standard deviation. Ninety-five of these unbonds (82.6 percent) were confirmed to be adhesive voids by x-radiography; however it must be noted that x-radiography can only detect missing adhesive mass, or voids, whereas acoustic methods detect missing adhesion.

This desirable result was highly dependent on using a common or identical reference specimen. Use of the same reference specimen provided uniform control of the sensitivity of the instruments, hence nearly uniform unbond detection and definition. Use of a lightly dissimilar reference specimen (the same in composition the geometry, but chosen from the 60th rather than 40th percentile rank in LBI) resulted in causing the area detected as unbonded to nearly double.

7.8 Conclusions

Quantitative NDE of adhesively bonded FRP lap joints for shear strength can be routinely accomplished by using an electrically modified, commercially available bond tester to obtain two NDI indicators. Obtained by a simple cost-effective inspection procedure, these indicators provide bond inspection data which show fairly good correlations with bond joint strength, and good sensitivity over the range of expected use.

Each of these two indicators has a specific bond NDI application. The first, LBI, indicates local bond integrity over a region approximately 2 cm2. The LBI is read from a meter wired into the bond tester circuitry. The second, BMF, is bond merit factor. It indicates regional bond integrity over a bond line ranging in light from about 10 to perhaps 40 cm. The BMF is computed from bond inspection data defining unbonded regions of the bond line. The definition of these unbonds is based upon the operational sensitivity of the bond tester which is initially adjusted during calibration with a bond reference specimen. Ultimately, the reliability and validity of the method is based upon the reference specimen selection and verification procedure.

7.9 Evaluation of implementation strategies

Both methods can be implemented in an on-site, in-line NDE methodology which is far more valuable than a product screening or batch inspection approach, often implemented after inferior or defective product has been detected or suspected in a production lot. An on-site NDE technique can evaluate the product immediately after production, thus limiting the number of defective products produced to populate the production line between the process and the inspection station, thus providing information for correcting the process almost immediately. This quality feed-back loop concept is shown schematically in Fig. 7-15, with the process at step 1, the product produced by it at step 2, and the evaluation of the product at step 3. This evaluation then provides early feed-back information in step 4 to control the process for production of the n+1 item being produced in step 1, where n is the number of items on the production line, conveyor, or in the "pipeline" between the evaluation and the resulting correction to the process. In this scheme, subsequent product would fall within the specification limits (SL) shown at the top of the figure. After the process is corrected and the correction "fixed", so that the correct production control parameters provide consistent quality, the routine NDE monitoring scheme shown in Fig.7-15 can be replaced by an auditing NDE scheme, or under certain "Deming inspection criteria" [14, 15], eliminated altogether, as shown in Fig. 7-16.

W. Edwards. Deming [15], Emmanuel P. Papadakis [16], Philip B. Crosby [17] and others have developed quantitative methods for establishing inspection criteria that quantify the value of inspection to the manufacturing enterprise when NDE is implemented in the appropriate production situations, and eliminated when the process is shown to be under control and capable, as illustrated by the sketch in Fig.7-16. At that point the benefit of inspection to the process is diminished, as well as its value to the manufacturing enterprise; hence its cost can no longer be justified, except for critical components, as addressed in the references cited above.

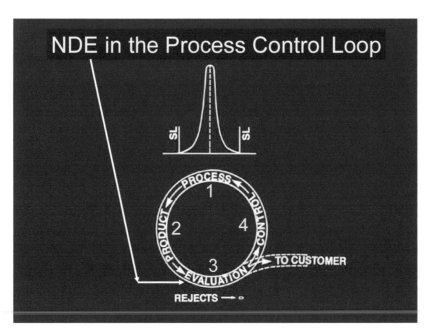

Fig. 7-15. Process quality control loop

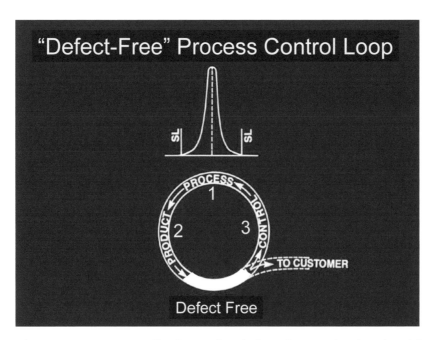

Fig. 7-16. Process quality loop after defect-free production is achieved

8. Concluding Summary and Value in Quality Improvement

8.1 The need for two complimentary adhesive bond NDE technologies

Both the high-frequency and low-frequency ultrasonic NDE technologies have valuable roles in assuring adhesive bond-joint quality in metal and polymer assemblies. Each of the ultrasonic NDE technologies discussed herein is complementary. As explained in sections 5.7.2 and 7.3, the high-frequency pulse-echo technique is effective only when there are significant differences in acoustic impedances between the substrate and adhesive materials. Since such is not the case when polymer composites and plastics are bonded with a polymer adhesive, another more effective NDE method is recommended in order to offer a complete set of NDE technologies for quality assurance of adhesive bonds in both of these materials commonly used in automotive and aerospace assemblies.

8.2 Adhesive bond joint NDE capabilities of 20 MHz pulse-echo technique

Attenuation in the successive echo amplitudes of reverberations in the substrate metal is a reliable indicator of the adhesion state (bonded or unbonded) at interface 1, and therefore can be used to detect adhesion at interface 1, even when there is no reflected echo from the adhesive-air interface, when unbonded, or at the adhesive-metal interface, when bonded at interface 2. Changes in the attenuation of the echoes reverberating in the metal, as measured by changes in the successive echo amplitude ratios, $A_n/A_{(n-1)}$, can also provide a reliable indicator of changes in the metallurgical history of the substrate. This could quite useful as a metal batch or allow change indicator.

The phase preference is negative when there is no bond at interface 2, but positive when a bond does exist there; therefore the phase of the echo reflected from interface 2 is a reliable indicator of the bond state at interface 2. The resolution capability is 4 mm or better on bonds in steel and aluminum sheet assemblies.

8.3 Adhesive bond joint NDE capabilities of 25 kHz Lamb wave technique

The 25 kHz Lamb wave technique has proven capable of detecting unbonded regions of a plastic or polymer composite bond joint as small as 1 cm. Probability of detecting adhesive voids varied from about 0.25 at the 1-cm limit of detection, to about 0.5 at 1.5 cm, to 0.95 at 2.0 cm. The probability of detecting unbonded adhesive ranged from about 0.25 at the same 1-cm limit of detection, to about 0.5 at 3.3 cm, to 0.95 at 4.0 cm. Therefore it can be said that adhesive voids larger than 2 cm and unbonds larger than 4 cm have a 95% probability of detection by this technique.

Although beyond the purview of this investigation, the ability to measure changes in Lamb-wave velocity, and have those changes identified by direction of propagation, lends the use of this technology to the determination of the prevailing direction of fiber orientation in long-fiber and chopped-fiber reinforced composites [7].

8.4 Advantages and limitations of these two complementary methods

The 20 MHz pulse-echo technique requires a liquid couplant, the 25 kHz technique does not. The 20 MHz method has a 4 mm resolution and the small transducer can be used in tight places. The 25 kHz technique does not detect 4mm unbonds well, but detects 3 cm unbonds quickly, when there is sufficient working room for the 42 mm-diameter probe along the bond joint. The 25 kHz method is effective on plastics and composites, where the acoustic impedance of the adhesive is close to that of the substrate. The 20 MHz method is not effective in such cases, but is far more effective on metal bond joints than the Lamb wave technique.

8.5 Novel NDE capabilities established

Contributions resulting from this research to develop improved ultrasonic NDE techniques offer advancements to the state of the art for NDE of adhesive bonds. These include

• Nondestructive detection and classification of bond states by being able to detect the state of adhesion in metal-adhesive bond joints, and distinguish whether the bond or unbond is at the near metal-adhesive interface, interface 1, or at the far adhesive-substrate interface, interface 2.

• Detect and distinguish kissing bonds (adhesive bonds that have very little strength) in polymer composite bond joints. The nondestructive identification of these weak bonds has posed a significant challenge to the adhesive bond NDE community because the NDE of bond joints for strength is not fundamentally directly achievable by ultrasonic pulse-echo or through-transmission techniques.

• Classification and detection of kissing unbonds (joints in which there is adhesive contact, but no strength) in both adhesively bonded metal and polymer composite joints.

• Analytical modeling of the propagation and reflection of longitudinal waves for pulse-echo simulations of A-scans of both adhesive bond states and at both adhesive-substrate interfaces, 1 and 2, over a practical range of adhesive and substrate materials and thicknesses. This modeling demonstrated the ineffectiveness of the ultrasonic pulse-echo approach to the NDE of bond joints in which the acoustic impedance differences between the substrate and adhesive are small. A through transmission NDE approach could be effective, but would require access to both sides of the bond joint. Since access to both sides of the bond joint is denied by constraints concomitant with implementations, the asymmetric Lamb-wave approach is recommended.

• Analytical modeling of asymmetric Lamb wave propagation along bonded joint interfaces with small impedance differences, and along unbonded joint interfaces with small impedance differences phase shift and amplitude increase as two unbond detection modes.

• Experimental results which verify the modeling simulations was acquired and reported

— by A-scans with longitudinal waves in pulse-echo NDE for amplitude-based and attenuation-based detection of bond state at interface 1,

— by A-scans with longitudinal waves in pulse-echo NDE for phase-based detection of bond state at interface 2,

— by velocity-based detection with Lamb waves for which measuring phase shifts to detect unbonds in polymer composites was demonstrated, and

— by amplitude-based detection with Lamb waves.

These two complementary ultrasonic NDE techniques provide a complete bond evaluation suite to nondestructively assure the quality of adhesive bonds in metal, plastic and polymer composite bond joints.

8.6 Remaining challenges, research and feasibility studies

The automation of the pulse-echo technique is under consideration, and the automation of the echo interpretation phase is already well along in development and has experienced plant trials. There are no plans to automate the Lamb-wave method. Work is currently underway to provide computer-automated interpretation of the echoes and provide an on-screen display of the bond-quality results.

Research and further developments in this area are continuing in order to make the ultrasonic pulse-echo approach operationally simple, more operator-friendly and the interpretation of the results less operator-dependent. This includes a reduction in the complexity of the instrumentation display to a simple auditable and/or graphic indicator. Feasibility studies have been undertaken in both the laboratory and in several manufacturing facilities to determine the manufacturing feasibility, and the degree of automation that would be appropriate to provide rapid, cost-effective implementation.

Manufacturing feasibility was demonstrated in several plants, using the operational simple instrumentation that had its display reduced to a simple auditable and graphic indicator by instrument producer and technical services corporation, Tessonics, Inc.. These studies showed that manufacturing process quality improvements could be achieved at reduced cost by having NDE inspectors detect unbounded regions before additional value-added manufacturing processes were invested in the assembly. Early detection of these defective bond joints also provided feedback information that could be promptly used to quickly correct the defective adhesive-bonding processing producing them.

These unbonds were sometimes caused by the adhesive application robot missing the specified bond line, or by insufficient adhesive being applied to make contact with both substrate surfaces. The latter may be due to pillowing of the sheet-metal between spot welds, lack of good bond-joint coordination between plastic or composite parts, lack of the proper amount of dispensed adhesive, adhesive and/or substrate bond interface surfaces pacified by contamination or surface curing, or excessive squeeze-out caused by a force applied on the bond, then removed before curing. Other causes of unbonds that are listed in chapter 1, section 1.2, are less commonly observed here.

An encouraging report of some manufacturing implementation efforts and results, using the Tessonics operator-friendly instrument, are reported in reference 203.

References

1.   Automotive Coatings, Sealants & Adhesives at http://www.mindbranch.com/products/R154-1552.html

2.   G.B. Chapman II, "A Nondestructive Method of Evaluating Adhesive Bond Strength in Fiberglass Reinforced Plastic Assemblies", *Joining of Composite Materials, ASTM STP 749*, pp. 32 – 60, K. T. Kedwards, ed., American Society for Testing and Materials, (1981).

3.   G.B. Chapman II, "Nondestructive Inspection Technology for Quality Assurance of Fiber-Reinforced Plastic Assemblies, SAE Transactions 91 (1982)

4.   G.B. Chapman II, "Ultrasonic Tests for Automotive Composites", *Nondestructive Testing Handbook*: Section 18, Ultrasonic Testing Applications in the Transportation Industries, pp. 669 – 693, Paul McIntire, editor. The American Society for Nondestructive Testing, Columbus, OH, (1991).

5.   American Society for Testing and Materials Standard D1002-01, Standard Test Method for Apparent Shear Strength of Single-Lap-Joint Adhesively Bonded Metal Specimens by Tension Loading (Metal-to-Metal), ASTM International.

6.   American Society for Testing and Materials Standard D3163-01, 1979, Standard Test Method for Determining Strength of Adhesively Bonded Rigid Plastic Lap-Shear Joints in Shear by Tension Loading, ASTM International.

7.   Gilbert B. Chapman II; "Quality Systems for Automotive Plastics"; Chapter 10 of **Composites Materials Technology – Processes and Properties**, pp. 351-389, Edited by P. K. Mallick and S. Newman, Hanser, (1990)

8.   R.D. Adams and B.W. Drinkwater. Int. J. Mater. Prod. Technol. **14**, 385 (1999).

9.   R.D. Adams and P. Cawley. NDT Int. **21**, 208 (1988).

10. I.J. Munns and G. A. Georgiou. Insight, **37**, 941 (1995).

11. G.B. Chapman II, "Polymer Composites for Improved Energy Efficiency in the Automotive Industry", **Proceedings of the Symposium on Physical Sciences and Advanced Vehicle Technologies**, Roman Gr. Maev, editor, Terento, ON, CA, (June 2000).

12. L.T. Drzal, A.K. Bhurke, M.J. Rich, P. Askeland; "Surface Treatment of Plastics with UV Light in Air to Improve Adhesion --- An Economical and Environmentally Benign Process." **Proceedings of the 8th International Coatings for Plastics Symposium** (2005)

13. Elena Yu. Maeva, Inna Severina, Sergiy Bondarenko, Gilbert Chapman, Brian O'Neill, Fedar Severin, and Roman Gr. Maev; "Acoustical Methods for the Investigation of Adhesively Bonded Structures: A Review", **Canadian Journal of Physics**, Vol. 82, No. 12, pp. 981-1025, (2004).

14. W.E. Deming, Quality, Productivity and Competitive Position, Massachusetts Institute of Technology (MIT), Center for Advanced Engineering Study, (1982)

15. W.E. Deming, **Out of the Crisis**; MIT (1986)

16. E.P. Papadakis, **Financial Justification of Nondestructive Testing: Coat of Quality in Manufacturing**; CRC Pr I Llc (August 2006)

17. P.B. Crosby, **Quality is Free**; McGrae-Hill, New York (1979)

18. J.C. Maxwell. Philos. Trans. R. Soc. **157**, 49 (1867).

19. R.B. Thompson and D.O. Thompson. J. Adhes. Sci. Tech. **5**, 583 (1991).

20. Briggs, Acoustic microscopy. Oxford University Press, New York. 1992.

21. R.G. Maev. Einsatz der Akustomicroskope in den Materialwisenschaften. Review of the BRD-USSR bilateral seminar "Microscopy in Material Sciences". Science Publishing, Moscow. 1988. p. 35.

22. R.G. Maev, J. H. Sokolowsky, H.T. Lee, E.Y. Maeva, and A.A. Denisov. Mater. Charact. **46**, 263 (2001).

23. B. Drinkwater and P. Cawley. Ultrasonics, **35**, 479 (1997).

24. M. Sutin, C. Delclos, and M. Lenclud. *In* Proceedings 2nd Int. Symp. on Acoustic and Vibration Surveillance Methods and Diagnostic Techniques. Senlis, France. 1995. p. 725.

25. A.M. Sutin and V. E. Nazarov. Radiophys. Quantum Electron. **38**, 109 (1995).

26. Y. Bar-Cohen, A. K. Mal, and C.-C. Yin. J. Adhesion, **29**, 257 (1989).

27. S. Rokhlin, M. Hefets, and M. Rosen. J. Appl. Phys. **52**, 841 (1981).

28. W. MacBain and D.G. Hopkins. J. Phys. Chem. **29**, 88 (1925).

29. M.C. van der Leeden and G. Frens. Adv. Eng. Mater. **4**, 280 (2002).

30. G. Fourche. Poly Eng. Sci. **35**(12), 957 (1995).

31. L.H. Lee. Adhesive bonding. Plenum Press, New York. 1991.

32. S.G. Abbot. Int. J. Adhes. Adhes. **5**, 7 (1985).

33. G. Kumar and K. Ramani. J. Comp. Mater. **34**, 1582 (2000).

34. M. Kazayawoko, J.J. Balatinecz, and L.M. Matuana. J. Mater. Sci. **34**, 6189 (1999)

35. D.E. Packhman and C. Johnson. Int. J. Adhes. Adhes. **14**, 131 (1994).

36. R.W. Messler and S. Gene. J. Thermoplast. Comp. Mater. **11**, 200 (1998).

37. N.H. Ladizesry and I.M. Ward. J. Mater. Sci. **24**, 3763 (1989).

38. S.S. Voyutski. Autohesion and adhesion of high polymers. Wiley-Interscience, New York. 1963.

39. H. Pravatareddy, J.G. Dillarad, J.E. McGrath, and D.A. Dillard. J. Adhes. **69**, 83 (1999).

40. C.M. Hansen and L. Just. Int. Eng. Chem. Res. **40**, 21 (2001).

41. R.M. Vasenin. Adhesion: fundamentals and practice. McLaren, London. 1969 p. 29.

42. P.G. de Gennes. J. Chem. Phys. **55**, 572 (1971).

43. M. Doi. Introduction to polymer physics. Clarendon Press, Oxford. 1995.

44. W.W. Graessley. Adv. Polymer Sci. **47**, 76 (1982).

45. B.V. Deryagin and V.P. Smilga. J. Appl. Phys. **38**, 4609 (1967).

46. S. Yang, L. Gu, and R.F. Gibson. Compos. Struct. **51**, 63 (2001).

47. D.A. Hays. *In* Fundamentals of adhesion. *Edited by* L.H. Lee. Plenum Press, New York. 1999. p. 249.

48. J.Q. Feng and D.A. Hays. Powder Technol. **135–136**, 65 (2003).

49. H. Zhou, M. Götzinger, and W. Peukert. Powder Technol. **135–136**, 82 (2003).

50. W.S. Czarnecki and L.B. Schein. J. Electrostat. **61**, 107 (2004).

51. D.A. Hays. J. Adhes. Sci. Technol. **9**, 1063 (1995).

52. Hays. *In* Particle on Surface: 1. Detection, adhesion and removal. *Edited by* K.L. Mittal. Plenum, New York. 1988, p. 223.

53. H. Mizes, M. Ott, E. Eklund, and D. Hays. Colloids Surfaces A: Physicochem. Eng. Aspects, **165,** 11 (2000).

54. L.H. Sharpe and H. Schonhorn. Adv. Chem. Series, **8**, 189 (1964).

55. W.A. Zisman. *In* Adhesion science and technology. *Edited by* L.H. Lee. Vol. A. Plenum Press, New York. 1975. p. 55.

56. K.L. Mittal. *In* Adhesion science and technology. *Edited by* L.H. Lee. Vol. A. Plenum Press, New York. 1975. p. 129.

57. T. Young. Philos. Trans. R. Soc. **95**, 65 (1805).

58. Duprée. Theorie mechanique de la chaleur. Gauthier-Villars, Paris. 1869. p. 393.

59. R. J. Good, M.K. Chaudhury, and C.J. van Oss. Fundamentals of adhesion. *Edited by* L.H. Lee. Plenum Press, New York. 1991. Chap. 4.

60. F. M. Fowkes. J. Adhesion Sci. Tech. **1**, 7 (1987).

61. D. A. Allara, F.M. Fowkes, J. Noolandi, G.W. Rubloff, and M.V. Tirrell. Mat. Sci. Eng. **83**, 213 (1986).

62. F. M. Fowkes. J. Phys. Chem. **67**, 2538 (1963).

63. D. K. Owens and R.C. Wendt. J. Appl. Polymer Sci. **13**, 1741 (1969).

64. S. Wu. Polymer interface and adhesion. Marcel Dekker, New York. 1982.

65. C. Della Volpe, D. Maniglio, M. Brugnara, S. Siboni, and M. Morra. J. Colloid Interface Sci. **271**, 434 (2004).

66. C. J. van Oss, R. J. Good, and M.K. Chaudhurry. Langmuir, **4**, 884 (1988).

67. C. J. van Oss and R. J. Good. Interfacial forces in aqueous media. Marcel Dekker, New York. 1994.

68. F. M. Fowkes. Surface and interfacial aspects of biomedical polymers. Vol. 1. *Edited by* D. Andrade. Plenum Press, New York. 1985. Chap. 9.

69. R.J. Good and C.J. van Oss. Modern approach to wettability: Theory and application. *Edited by* M.E. Schrader and G. Loed. Plenum Press, New York. 1991. Chap. 1.

70. M. Morra. J. Colloid Interface Sci. **182**, 312 (1996).

71. J.C. Berg (*Editor*). Wettability. Surfactant Science Series. Dekker, New York. 1993. p. 49.

72. C. Della Volpe and S. Siboni. J. Colloid Interface Sci. **195**, 121 (1997).

73. M. Connor, J.-E. Bidaux, and J.-A.E. Manson. J. Mater. Sci. **32**, 5039 (1997).

74. L.S. Penn and E. Defex. J. Mater. Sci. **37**, 505 (2002).

75. J.J. Bikerman. The Science of adhesive joints. 2nd ed. Academic Press, New York. 1968.

76. F. Fabulyak. Molecular flexibility in the border layers. Naukova Dumka, Kiev. 1983.

77. R.A. Veselovsky. Adhesion of polymers. McGraw-Hill, New York. 2001.

78. R.G. Good. Measurement of adhesion of thin film, thin film and bulk coating. *Edited by* K.L. Mittal. ASTM **STP 640**, 18 (1978).

79. M. Kalnins and J. Ozolins. Intl. J. Adhes. Adhes. **22**, 179 (2002).

80. Perepechko, Acoustic Methods of Investigating Polymers, Mir Publishers, Moscow, (1973) 1975, p.5.

81. J. Szilard (*Editor*). Physical principles of ultrasonic testing in ultrasonic testing. John Willey & Sons, New York. 1982. pp. 1–24.

82. D.E. Bray and D. McBride (*Editors*). Acoustic testing of materials *In* Nondestructive testing techniques. John Wiley & Sons, Inc. New York. 1992. pp. 253.

83. R.G. Maev, H. Shao, and E. Yu. Maeva. Mater. Charact. **41**, 97 (1998).

84. L. Goglio and M. Rossetto. NDT&E Intl. **32,** 323 (1999).

85. L. Goglio and M. Rossetto. Ultrasonics, **40,** 205 (2002).

86. R.E. Challis, R.G. Freemantle, G.P. Wilkinson, and J.D.H. White. Ultrasonics, **34,** 315 (1996).

87. W.J. Thompson. J. Appl. Phys. **21,** 89 (1950).

88. N.A. Haskell. J. Bull. Seism. Soc. Am. **43**, 17 (1953).

89. D.S.G. Pollock. A handbook of time-series analysis, signal processing and dynamics. Academic Press, London. 1999. p. 130.

90. K. Vine, P. Cawley, and A.J. Kinloch. NDT&E Intl. **35**, 241 (2002).

91. A.K. Moidu, A.N. Sinclair, and J.K. Spelt. Res. NDE, **11**, 81 (1999).

92. Pilarski and J.L. Rose. J. Appl. Phys. **63**(2), 300 (1998)

93. H.G. Tattersall. J. Appl. Phys D: Appl. Phys. **6**, 819 (1973).

94. A.V. Clark, Jr. and S.D. Hart. Mater. Eval. **40**, 866 (1982).

95. D.N. Sinha. Acoustic resonance spectroscopy (ARS), IEEE Potentials, **11**, 10 (1992).

96. B. Zadler, J.H.L. Le Rousseau, J.A. Scales, and M.L. Smith. Geophys. J. Int. **156**, 154 (2004).

97. A.F. Brown. *In* Ultrasonic testing. *Edited by* J. Szilard. John Wiley & Sons, New York. 1982. pp. 167.

98. R.G. Leisure and F.A. Willis. J. Phys. Condens. Matter, **9**, 6001 (1997).

99. Migliori and T.W. Darling. Ultrasonics, **34**, 473 (1996).

100. J. Szilard. *In* Ultrasonic testing. *Edited by* J. Szilard. John Wiley & Sons, New York. 1982. pp. 36.

101. R.B. Schwarz and J.F. Vuorien. J. Alloys Comp. **310**, 243 (2000).

102. C.C.H. Guyott, P. Cawley, and R.D. Adams. J. Adhesion, **20**, 129 (1986).

103. D.J. Hagemaier. Non-Destruct. Test. **5**, 38 (1972).

104. Yu.V. Lange. Sov. J. Nondestr. Test. (Eng. Trans.) **12**, 5 (1976).

105. P. Cawley. IEEE Ultrasonics Symposium, 1992, p. 767.

106. P. Cawley and T. Pialucha. IEEE Ultrasonics Symposium, 31 Oct.–3 Nov. 1993. p. 729.

107. W. Wang and S. Rokhlin. J. Adhes. Sci. Technol. **5**(8), 647 (1991).

108. L. Adler, S. Rokhlin, and A. Baltazar. IEEE Ultrasonic Symposium, 2001, Vol. 1. pp. 701.

109. S. Yang, L. Gu, and R.F. Gibson. Compos. Struct. **51**, 63 (2001).

110. A.M. Robinson, B.W. Drinkwater, and J. Allin. NDT&E Intl. **36**, 27 (2003).

111. R. Kašis and L. Svilainis. Ultrasonics, **35**, 367 (1997).

112. S.H. Diaz Valdes and C. Soutis. J. Sound Vib. **228**(1), 1 (1999).

113. G.J. Curtis. *In* Ultrasonic testing. *Edited by* J. Szilard. John Wiley & Sons, New York. 1982. p. 495.

114. P. Cawley. NDT Technology in aerospace, IEE Colloquium, 15 Jan. 1990. p. 6.

115. T.M. Whitney and R.E.Gr. Green. Ultrasonics, **34**, 347 (1996).

116. D. Lescribaa and A. Vincent. Surf. Coat. Technol. **81**, 297 (1996).

117. C.F Quate. Phys. Today, **38**(8), 34, (1985).

118. B. O'Niell and R.Gr. Maev. Can. J. Phys. **78**, 803 (2000).

119. Y. Zheng, R.Gr. Maev, and I.Yu. Solodov. Can. J. Phys. **77**, 927 (1999).

120. R. Brown. Handbook of polymer testing. Marcel Dekker, Inc., New York. 1999.

121. D.J. Hagemaier. *In* ASM Handbook nondestructive evaluation and quality control. Vol. 17. 9th ed. USA. 1996. p. 610.

122. R.A. Lemons and C.F. Quate. Acoustic Microscopy, Physical Acoustics, Edited by W.P. Mason and R.N. Thurston; Academic Press, New York, vol 14, pp

123. Cros, V. Gigot, and G. Despaux. Appl. Surf. Sci. **119**, 242 (1997).

124. K. Yamanaka, Y. Nagata, T. Koda, and K. Karaki. *In* Ultrason. Symp. Proc., Vol. 2, (Cat. #90CH2938-9), IEEE, Piscataway, NJ, USA. 1990. pp. 913.

125. B. Hadimioglu and J.S. Foster. J. Appl. Phys. **56**, 1976 (1984).

126. D.R. Billson and D.A. Hutchins. Brit. J. NDT, **35**, 705 (1993).

127. B. Drinkwater and P. Cawley. Brit. J. NDT, **36**, 430 (1994).

128. B. Drinkwater and P. Cawley. Mater. Eval. **55**, 401 (1997).

129. D.W. Schindel, D.A. Hutchins, and W.A. Grandia. Ultrasonics, **34**, 621 (1996).

130. T. Gudra, M. Pluta, and Z. Kojro. Ultrasonics, **38**, 794 (2000).

131. E. Blomme, D. Bulcaen, and F. Declercq. NDT&E Intl. **35**, 417 (2002).

132. B. Cross, M.F. Villat, and F. Augereau. J. Mater. Sci. **32**, 2655 (1997).

133. H. Lamb. Proc. R. Soc. London, **XCIII**, 114 (1917).

134. S.I. Rokhlin and D. Marom. J. Acoust. Soc. Am. **80**, 585 (1986).

135. Y. Bar-Cohen and D.F. Chimenti. Rev. Prog. Quantum. NDE, **5B**, 1199 (1985).

136. J.D. Achenbach. Int. J. Solids Struct. **37**, 13 (2000).

137. P.A. Fomitchov, S. Krishnaswamy, and J.D. Achenbach. Optics Laser Technol. **29**, 333 (1997).

138. A.K. Mal, P.C. Xu, and Y. Bar-Cohen. Int. J. Eng. Sci. **27**, 779 (1989).

139. Pilarsky, J.L. Rose, and K. Balasubramaniam. J. Acoust. Soc. Am. **82**, 21 (1987).

140. P.B. Nagy and L. Adler. J. Appl. Phys. **66**, 4658 (1989).

141. L. Singher. Ultrasonics, **35**, 305 (1997).

142. R.Y. Vasudeva and G. Sudheer. Int. J. Solids Struct. **39**, 559 (2002).

143. S.C. Cowin and J.W. Nunziato. J. Elasticity, **13**, 125 (1983).

144. K. Heller, L.J. Jacobs, and J. Qu. NDT&E Intl. **33**, 555 (2000).

145. D.W. Schindel, D.S. Forsyth, D.A. Hutchins, and A. Fahr. Ultrasonics, **35**, 1 (1997).

146. NDE Resource Center at Iowa State University, http://www.ndt-ed.org/EducationResources/CommunityCollege/Ultrasonics/Introduction/history.htm

147. I.Yu. Solodov and C. Wu. Acoust. Phys. **39**, 476 (1993).

148. I.Yu. Solodov. Ultrasonics, **36**, 383 (1998).

149. P.B. Nagy. J. Nondestr. Eval. **11**, 127 (1992).

150. E.L. Chez, J. Dundurs, and M. Comninou. Int. J. Solids Struct. **19**, 579 (1983).

151. M. Comninou and J. Dundurs. Proc. R. Soc. London. A, **368**, 141 (1979).

152. J. Richardson. Int. J. Eng. Sci. **17**, 73 (1979).

153. W. Yue-Sheng, Y. Gui-Lan, and G. Bing-Zheng. Int. J. Solids Struct. **35**, 2001 (1998).

154. B. O'Neill, R.Gr. Maev, and F. Severin. *In* Review of progress in quantitative nondestructive evaluation. Vol. 20B. AIP, Mellville, New York. 2000. p. 1261.

155. Baltazar, S.I. Rokhlin, and C. Pecorari. J. Mech. Phys. Solids, **50**, 1397 (2002).

156. O.V. Rudenko and C.A. Vu. Acoust. Phys. **40**, 593 (1994).

157. S. Hirose and J.D. Achenbach. J. Acoust. Soc. Am. **93**, 142 (1993).

158. N. Yoshioka. Tectonophysics, **277**, 29 (1997).

159. L.J. Pyrak-Nolte, L.R. Myer, N.G.W. Cook. J. Geophys. Res. **95**, 8617 (1990).

160. R.B. Thompson and J.M. Baik. J. Nondestr. Eval. **4**, 177 (1984).

161. C. Pecorari and P. Kelly. *In* Review of progress in quantitative nondestructive evaluation. Vol. 18. AIP, Mellville, New York. 1999. pp. 1471.

162. J.D. Achenbach and O.K. Parikh. J. Adhes. Sci. Technol. **5**, 601 (1991).

163. D.D. Palmer, D.K. Rehbein, J.F. Smith, and O. Buck. J. Nondestr. Eval. **7**, 164 (1988)

164. Baltazar, S.I. Rokhlin, and C. Pecorari. J. Mech. Phys. Solids, **50**, 1397 (2002).

165. Lavrentyev and S.I. Rokhlin. J. Acoust. Soc. Am. **103**, 657 (1998).

166. J.D. Achenbach, O.K. Parikh, and D.A. Sotiropoulos. *In* Review of progress in quantitative nondestructive evaluation. Vol. 8. *Edited by* D.O. Thompson and D.E. Chimenti. 1989. p. 1401.

167. Lavrentyev and J.T. Beals. Ultrasonics, **38**, 513 (2000).

168. D.D. Palmer, D.K. Rehbein, J.F. Smith, and O. Buck. J. Nondestr. Eval. **7**, 153 (1988).

169. O. Buck, D.K. Rehbein, R.B. Thompson, D.D. Palmer, and L.J.H. Brasche. *In* Review of progress in nondestructive evaluation. Vol. 8. *Edited by* D.O. Thompson and D.E. Chimenti. 1989. pp. 1949.

170. V.F. Humphrey. Nonlinear propagation in ultrasonic fields: measurements, modeling and harmonic imaging. Ultrasonics, **38**, pp. 267-272 (2000).

171. Handbook of Chemistry and Physics, CRC Press, 63rd Edition

172. MatWeb, a source of online materials information at http://www.matweb.com/search/GetKeyword.asp

173. Wikipedia at http://en.wikipedia.org/wiki/Acoustic_wave_equation

174. A.A. Denisov. *Modeling and Optimization of Non-Phased 2D Ultrasonic Arrays*. Physics Ph.D. Dissertation, University of Windsor, Windsor, On, CA

175. L.W. Schmerr, Jr; *Fundamentals of Nondestructive Evaluation, A modeling Approach*, Plenum Press, New York, ISBN 0-306-45752-0, 1988

176. P. Palanichamy, A. Joseph, T. Jayakumart, and R. Baldev; *NDT & E International* (NDT E int.) ISSN 0963-8695 1995, vol. 28, no3, pp. 179-185

177. J. F. Bussiere, M. Dubois, A. Moreau and J.-P. Monchalin; "Characterizing Materials with Laser Ultrasonics", in *Nondestructive Characterization of Materials IX*, edited by R. E. Green, Jr, American Institute of Physics, pp 131-141, 1999

178. A.N. Diógenes, E.A. Hoff, and C.P. Fernandes, "Grain size measurement by image analysis: An application in the ceramic and in the metallic industries" ; Proceedings of COBEM 2005 18th International Congress of Mechanical Engineering Copyright © 2005 by ABCM November 6-11, 2005, Ouro Preto, MG

179. K. Goebbels, *Materials Characterization for Process Control and Product Conformity*, CRC Press, Boca Raton (1994)

180. M. Moshfeghi and P.D. Hanstead "Ultrasound Reflection of Cracks in Polyester Resin; *Polymer NDE*, K.G.H. Ashby, Editor, Technomic, pp 275-305 (1984)

181. R.W. Morse, "Ultrasonic Attenuation in Metals by Electron Relaxation", Royal Society Mond Laboratory, Cambridge University, Cambridge, England (1955)

182. Panametrics 20 MHz Transducer Response Characteristics Data Sheet

183. R.Gr. Maev, S.A. Titov and G.B. Chapman, "Evaluation of adhesive bonds in sheet-metal assembles by a 20 MHz ultrasonic pulse-echo technique"; British Institute of Nondestructive Testing, in Statford-upon-Avon, United Kingdom. (September 2006)

184. Iowa State University NDE Resource Center, http://www.ndt-ed.org/EducationResources/CommunityCollege/Ultrasonics/Physics/modepropagation.htm

185. University of Arkansas NDT Resource Center at http://www.ndt-ed.org/index_flash.htm

186. D.R. Bland, *Wave Theory and Applications*; Oxford University Press, New York (1988)

187. E. Yu. Maeva, I. Severina, and G.B. Chapman II, "Analysis of the Degree of Cure and Cohesive Properties of the Adhesive in a Bond Joint by Ultrasonic Techniques," *Research in Nondestructive Evaluation*, **Vol. 18,** Taylor & Francis (2006)

188. C. Pindinelli, et al. Macromol. Symp. **180**: 73 (2002).

189. N. T. Nguyen, M. Lethiecq, and J. F. Gerard. Ultrasonics, **33**: 323 (1995)

190. A. Maffezzoli, E. Quarta, A. M. Luprano, G. montagna, L. Nicolais. J. Appl. Pol. Sci. **73**: 1969 (1999)

191. F. Loinetto, R. Rizzo, V. A. M. Luprano, A. Maffezzoli, Material Sci. Eng. A. **370**: 284 (2004).

192. I. Alig, K. Nanke, G. P. Johary, J. Pol. Sci. Part B: Pol. Phys. **32**: 1465 (1994).

193. R. J. Freemantle, R. E. Challis, Meas. Sci. Technol. **9**: 1291 (1998),.

194. J. Sanjuan and M.A. Llorente, "Experimental analysis of the molecular theory of rubber elasticity", Journal of Polymer Science, Part B: Polymer Physics, vol 26, issue 2, pp.235-244 (1987 and 2003)

195. C. W. Macosko and D. R. Miller. Macromolecules. **9**(2): 199 (1976).

196. D. R. Miller and C. W. Macosko. Macromolecules. **9**(2): 206 (1976).

197. K. J. Rienks, "Introduction to the Fokker Bond Tester System" Fokker-VFW Report R-1983, Netherlands Aircraft Factories, The Netherlands

198. Ford Laboratory Test Method BU 17-1, Ford Motor Company

199. G.B. Chapman II and W.A. Gordon, Applied Spectroscopy, vol. 32, No. 1, 1978, pp. 46-53

200. American Society for Testing and Materials Standard, D 3165-73 (1979), ASTM International.

201. American Society for Testing and Materials Standard, D-790-71 (1978), ASTM International

202. D.F. Smith and C.V. Cagle, Applied Polymer Symposium, No. 3, 1966, pp. 411-434

203. G. B. Chapman, S. Titov, "Why Stick with Adhesives when Joint Quality is Required" Proceedings of the Pressure-Sensitive Tape Council (PSTC) Tech-33 Conference and Expo in Las Vegas, NV, May 12 – 14, 2010.

Related Publications

1. Meyer, Fred J. and Chapman, Gilbert B. II, "Nondestructive Testing of Bonded FRP Assemblies in the Auto Industry", <u>Adhesive Age</u>, Vol. 23, No. 4, April 1980, pp. 21 – 25.

2. Chapman, Gilbert B. II, Papadakis, Emmanuel P,. and Meyer, Fred S., "A Procedure for Nondestructive Inspection of Adhesive Bonds in Fiberglass Reinforced Plastic Assemblies", Ford Scientific Research Report SR 80-85, July 1980.

3. Chapman, Gilbert B. II, Ford Motor Company's Corporate-wide Adhesive Bond Inspection Method, Published as <u>Ford Laboratory Test Method (FLTM) BU 17-1</u>, July 1980.

4. Chapman, Gilbert B. II, "A Nondestructive Method of Evaluating Adhesive Bond Strength in Fiberglass Reinforced Plastic Assemblies", Ford Scientific Research Report SR 82-12, February 1982 and <u>Joining of Composite Materials</u>, <u>ASTM STP 749</u>, K. T. Kedwards, ed., American Society for Testing and Materials, 1981, pp. 32 – 60.

5. Chapman, Gilbert B. II, "Nondestructive Inspection for Quality Assurance of Fiber Reinforced Plastic Assemblies", Ford Scientific Research Report SR 82-17, February 1982 and <u>Society of Automotive Engineers Transactions</u>, SAE Warrendale Vol. 91, February 1982.

6. Papadakis, Emmanuel P., Chapman, Gilbert B. II, and Milenkovic, Val, "Computer Method for Linear Regression Analysis Applicable to Situations with Errors in Both Variables", Ford Scientific Research Report SR 82-142, October 1982.

7. Chapman, Gilbert B., Strength Quality Assurance for Automotive FRP Assemblies", <u>Proceedings of the Society of Manufacturing Engineers Composites in Manufacturing Conference</u>, Anaheim, CA, January 1983.

8. Papadakis, E. P., Bartosiewicz, L., Altstetter, J. D., and Chapman, G. B. II, "Morphological Severity Factor for Graphite Shape in Cast Iron and Its Relation to Ultrasonic Velocity and Tensile Properties", Ford Scientific Research Report SR 82-159, December 1982 and <u>American Foundrymen's Society Transactions</u>, Vol. 83-102, 1983.

9. Chapman, Gilbert B. II, "Nondestructive Evaluation for Quality Assurance of Plastic Assemblies" <u>Proceedings of the Second Annual Michigan State University Workshop on Advanced Materials – Recent Advances in Nondestructive Evaluation</u>, Traverse City, MI, 9 – 11 September 1984.

10. Chapman, Gilbert B. II, Papadakis, Emmanuel P,. and Meyer, Fred S., "A Nondestructive Inspection Procedure for Adhesive Bonds in FRP Assemblies", <u>Journal of the American Society of Body Engineers</u>, Vol. 12, No. 2, Fall 1984.

11. Chapman, Gilbert B. II, "The Cause and Cost of Quality in Joining of Materials", Proceedings of the Fifth Annual Michigan State University Workshop on Advanced Materials, Traverse City, MI, 9 – 11 September 1984.

12. Chapman, Gilbert B. II, "QSIT Enhances Composites Quality and Cost", Manufacturing Engineering, November 1987.

13. Chapman, Gilbert B. II, and Adler, Laszlo, "Nondestructive Inspection Technology in Quality Systems for Automotive Plastics and Composites", Society of Automotive Engineers Technical Paper No. 880155 and Society of Automotive Engineers Transactions, SAE, Warrendale, February – March 1988.

14. Adler, L., Rohklin, S., He, F., and Chapman, G., "Ultrasonic Shear Wave and Lamb Wave Amplitude Measurements in Adhesively Bonded Steel Plates", Proceedings of the Workshop on NDE of Adhesive Bond Strength, Orlando, FL, 13 – 14 April 1988.

15. Chapman, Gilbert B. II, and Lee, Pamela F., "The Development of a Knowledge-Based System for the Nondestructive Inspection of Composites", Society of Automotive Engineers Technical Paper No. 890246 and Society of Automotive Engineers Transactions, SAE, Warrendale, 1989.

16. Chapman, Gilbert B. II, "Quality Systems for Automotive Plastics", Chapter 10 in book Composite Material Technology – Processes and Properties, edited by P. K. Mallick and S. Newman. Hanser Publishers, Vienna and New York, 1990, pp. 349 – 393.

17. Chapman, Gilbert B. II, "Ultrasonic Tests for Automotive Composites", Nondestructive Testing Handbook: Section 18, Ultrasonic Testing Applications in the Transportation Industries, Paul McIntire, editor. The American Society for Nondestructive Testing, Columbus, OH, 1991, pp. 669 – 693.

18. Papadakis, Emmanuel P. and Chapman, Gilbert B. II, " Quantitative Nondestructive Evaluation of Adhesive Lap Joints in Sheet Molding Compound by Adaptation of a Commercial Bond Tester", International Advances in Nondestructive Testing, Warren J. McGonnagle, editor; Gordon and Breach, New York, 1991.

19. Chapman, Gilbert B. II and Hagerman, Edward M., "Nondestructive Inspection Technology for Quality Assurance of Automotive Composites", (Award winner) 9th Annual ASM/ESD Advanced Composites Technologies Conference Proceedings, ESD – The Engineering Society, Dearborn, MI, 1993.

20. Papadakis, Emmanuel P. and Chapman, Gilbert B. II, "NDE of Sheet Molding Compound Fiber Flow Failures", Review of Progress in Quantitative Nondestructive Evaluation, edited by Donald O. Thompson and Dale Chimenti, Vol. 13, Plenum Press, New York, 1993

21. Chapman, Gilbert B. II, "An Organization and Requirements for the Development of Polymer Composites for Automotive Structural Applications" Proceedings of The Twelfth International Annual Meeting of the Polymer Processing Society, Sorrento, Italy, May 1996.

22. Chapman, Gilbert B. II, "Chairman's Opening Overview and Introduction" <u>Proceedings of the Program of the Materials Conference of the 29th International Symposium on Automotive Technology and Automation (ISATA),</u> Florence, Italy; June 1996

23. Chapman, Gilbert B. II, "Nondestructive Inspection Technology for Quality Assurance in Fuel Cell Manufacturing" <u>Proceedings of the Fuel Cell Manufacturing Workshop</u>, Dearborn, Michigan, July 1996.

24. Chapman, Gilbert B. II, "The Role of Nondestructive Inspection Technology in Improving the Quality of Automotive Processes and Products", <u>Proceedings of the SPIE Conference on Nondestructive Evaluation for Process Control in Manufacturing</u>, SPIE Paper No. 2948-27, Scottsdale AZ, December 1996.

25. Chapman, Gilbert B. II, "The Need and Requirements for Polymer Composites in Automotive Structures", <u>Proceedings of the 24th Annual WATTEC Conference</u> in Knoxville, TN, by Oak Ridge National Laboratories, 24 February 1997.

26. Chapman, Gilbert B. II, "Some Automotive Applications of Infrared Thermography for Quality Improvements" <u>Proceedings of the Thermosense XIX: An International SPIE Conference on Thermal Sensing and Imaging Diagnostic Applications, SPIE Proceedings Vol.3056</u>; Orlando, Florida; April,1997.

27. Chapman, Gilbert B. II, "The Need and Requirements Driving The Development of Polymer Composites for Automotive Applications" <u>Proceedings of the 11th International Conference on Composite Materials (ICCM)</u>, Gold Coast, Australia, July 1997.

28. Nagesh, Surgesh; Patterson, Craig and Chapman, Gilbert B. II, "Analytical Correlations of Physical Tests on Typical Automotive Cross-Sections Made of Fiber-Reinforced Plastics", <u>Proceedings of The Materials for Energy-Efficient Vehicles Conference of the 31st International Symposium on Automotive Technology and Automation (ISATA),</u> Dusseldorf, Germany; June 1998.

29. Chapman, Gilbert B. II, "Automotive Applications of Polymer Composites: Cost, Quality, and Manufacturing Requirements" <u>Proceedings of the Fifth International Conference on Composites Engineering</u> in Las Vegas, Nevada, 10 July 1998.

30. Chapman, Gilbert B. II, "Fiscal and Functional Electron-Beam Curing Requirements for Enhancing the Properties of Polymeric Materials in Automotive Applications", <u>Proceedings of the Oak Ridge National laboratory's Third Annual Workshop on Electron-Beam Curing, Oak</u> Ridge Tennessee; April 1999.

31. Nagesh, Suresh and Chapman, Gilbert B. II, "Analytical Correlations of Physical Tests on Automotive Sections of Foam-Filled Fiber-Reinforced Plastics" <u>Proceedings of The Materials for Energy-Efficient Vehicles Conference of the 32nd International Symposium on Automotive Technology and Automation (ISATA),</u> Vienna, Austria; June 1999

32. Chapman, Gilbert B. II, "Emerging Automotive Applications of Thermoplastic Composites", <u>Proceedings of the 3rd International Conference on New Products and Production Technologies for A New Textile Industry"</u> Gent, Belgium, July 1999.

33. Chapman, Gilbert B. II, "A Thermoplastic Approach to A Composite Automotive Body", Proceedings of The International Body Engineering Conference (IBEC), SAE Paper, Detroit, MI, September 1999.

34. Chapman, Gilbert B., II and Vesey, Donald A. "Opportunities for Reinforced Plastics in Automotive Applications", Proceedings of the SPE's ANTEC 2000 Annual Technical Conference, Orlando, FL, 7-11 May 2000.

35. Nagesh, Suresh, Chapman, Gilbert B. II, and Oswald, Lawrence J., "Analytical Modeling of Automotive Sections of Foam-Filled Fiber-Reinforced Plastics", Materials for Energy-Efficient Vehicles Proceedings of the SPE ANTEC 2000 Annual technical Conference, Orlando, FL, 7-11 May 2000.

36. Chapman, Gilbert B. II, "Polymer Composites for Improved Energy Efficiency in the Automotive Industry", Proceedings of the Symposium on Physical Sciences and Advanced Vehicle Technologies, Roman Gr. Maev, editor, Toronto, ON, Canada, June 2000.

37. Chapman, Gilbert B. II, and Oswald, Lawrence J., "Need and Requirements for Adhesive Bonding of Thermoplastic Composites in Automotive Body Assemblies", Proceedings of the DaimlerChrysler Technology Colloquium on Structural Adhesive Bonding, Haus Lammerbuckel, Germany, 20 – 21 September 2000.

38. Nagesh, Suresh; Newaz, Golam W.; Chapman, Gilbert B. and Patterson, Craig, "Effect of Temperature and Loading Rate on Adhesively Bonded Fiber Reinforced Plastic Automotive Sections", Proceedings of the 2000 International Body Engineering Conference, Paper 2000-01-2730, 3 – 5 October 2000.

39. Chapman, Gilbert B. II, "Reinforced Thermoplastic Composites for Automotive Body Applications", Proceedings of the Society of Manufacturing Engineers (SME)/Composites Manufacturing Association (CMA) Composites Manufacturing and Tooling 2001 Conference. Orlando, FL, 21 – 23 February 2001.

40. Suresh, N., Newaz, Golam, Chapman, Gilbert B. II, Patterson, Craig, and Oswald, Lawrence J., "Characterization of Adhesive Failure and Modeling for Dynamic Analysis", Proceedings of the ANTEC 2001 Conference. Society of Plastic Engineers (SPE), Dallas, TX 6 – 10 May 2001.

41. Chapman, Gilbert B. II, "A Low-Cost Approach to A High-Volume Automotive Composites Body" Proceedings of the Plastics in Automotive Exteriors Conference in Frankfurt, Germany, 4 December 2001.

42. Chapman, Gilbert B. II, "A 25 kHz Acoustic Approach to the Nondestructive Evaluation of Adhesive Bond Joints in Automotive Assemblies and a Summary of Needs and Requirements for Inspection Improvements" ? Proceedings of the Canadian Association of Physicists Annual Congress 2002 ? in Quebec City on 3 June 2002.

43. Chapman, Gilbert B. II, "Five Familiar Factors Retarding the Growth of Composites in Automotive Applications", Proceedings of the 14th International Conference on Composite Materials (ICCM-14) Society of Manufacturing Engineers. San Diego, CA on 16 July 2003.

44. Gilbert B Chapman, II, "Infrared Monitoring of Friction Welds and Adhesive Bond Curing in Automotive Manufacturing", <u>Proceedings of the 16th World Conference on Nondestructive Testing</u>, Montreal, Canada, 30 August – 3 September 2004.

45. Gilbert B. Chapman, II, "Needs and Requirements Driving the Implementation of Nondestructive Inspection Technologies in Automotive Applications", <u>Proceedings of the 16th World Conference on Nondestructive Testing</u>, Montreal, Canada, 30 August – 3 September 2004.

46. Elena Maeva, Inna Severina, Sergiy Bondarenko, Gilbert B. Chapman, Brian O'Neill, Fedar Severin, and Roman Gr. Maev, "Acoustical Methods for the Investigation of Adhesively Bonded Structures: A Review", <u>Canadian Journal of Physics</u>, Vol. 82, No. 12, 2004, pp. 981-1025.

47. Gilbert B. Chapman, II, "Infrared Monitoring of Friction Welds and Adhesive Bond Curing in Automotive Manufacturing", <u>CINDE Journal Canada's National NDT Magazine</u>, Canadian Institute for NDE, vol. 26, no.3, May/June 2005.

48. R.Gr. Maev, S.A. Titov and G.B. Chapman, "Evaluation of adhesive bonds in sheet-metal assembles by a 20 MHz ultrasonic pulse-echo technique"; British Institute of Nondestructive Testing, in Statford-upon-Avon, United Kingdom. (September 2006)

49. Gilbert B. Chapman, Sadler, J, Maev, R.G., Titov, S., Maeva, E.Y., Severina, I. and Severin, F.; "Ultrasonic Pulse-Echo NDE of Adhesive Bonds in Sheet-Metal Assemblies", <u>Proceedings of the IEEE Ultrasonics Symposium</u>, Vancouver, BC, CA, 3 – 6 October 2006.

50. Ina Severina, Gilbert B. Chapman, Jeff Adler, Sergey Titov, Elena Maeva, Fedar Severin and Roman Maev; High Frequency Acoustic Imaging Methods for Adhesive Bond Microstructure Study and Physical, Chemical and Micromechanical Properties Evaluation, to the American Society for Nondestructive Testing Conference on Automotive Industry Advances with NDT, Dearborn, MI, 16 May 2007

51. Gilbert B. Chapman, and Sergey A. Titov, "Why Stick with Adhesives when Joint Quality is Required" Proceedings of the Pressure-Sensitive Tape Council (PSTC) Tech-33 Conference and Expo in Las Vegas, NV, May 12 – 14, 2010.

Printed in the United States
By Bookmasters